T0135066

The series "Studies in Computational Intelligence" (SCI) publishes new developments and advances in the various areas of computational intelligence—quickly and with a high quality. The intent is to cover the theory, applications, and design methods of computational intelligence, as embedded in the fields of engineering, computer science, physics and life sciences, as well as the methodologies behind them. The series contains monographs, lecture notes and edited volumes in computational intelligence spanning the areas of neural networks, connectionist systems, genetic algorithms, evolutionary computation, artificial intelligence, cellular automata, self-organizing systems, soft computing, fuzzy systems, and hybrid intelligent systems. Of particular value to both the contributors and the readership are the short publication timeframe and the world-wide distribution, which enable both wide and rapid dissemination of research output.

Indexed by SCOPUS, DBLP, WTI Frankfurt eG, zbMATH, SCImago.

All books published in the series are submitted for consideration in Web of Science.

More information about this series at https://link.springer.com/bookseries/7092

Cezary Biele

Human Movements in Human-Computer Interaction (HCI)

Cezary Biele
National Information Processing Institute
National Research Institute
Warsaw, Poland

ISSN 1860-949X ISSN 1860-9503 (electronic)
Studies in Computational Intelligence
ISBN 978-3-030-90006-9 ISBN 978-3-030-90004-5 (eBook)
https://doi.org/10.1007/978-3-030-90004-5

This Springer imprint is published by the registered company Springer Nature Switzerland AG
The registered company address is: Gewerbestrasse 11, 6330 Cham, Switzerland

Contents

Chapter 1
Introduction

1.1 General Remarks

This book is generally concerned with the human–computer interaction (HCI, for short) which has become over the last years or decades an object of interest, study and even implementations of immense importance. The main reason is that it is commonly believed that an interactive—in a synergistic cooperation and collaboration between the human being and "machine" (computer)—solution of complex problems we face in the present world is the most promising paradigm.

Human–computer interaction (HCI), which has flourished since the 1980s, is a multidisciplinary field of study which aims at developing tools and techniques, to be then implemented in software and hardware, which would make it possible to attain an effective and efficient interaction between the humans (the users) and computers. Needless to say that though in HCI emphasis in the beginning was on the computers, it had recently moved to other aspects, notably taking into account the very specific features of the human being who exhibits many inconsistencies like changeable preferences, intentions, judgements, can get tired, may behave in a strange way, etc.

Therefore, for the last years HCI has more and more been including results from other sciences exemplified by the cognitive sciences, psychology, etc.

To be more specific, in recent years, one could clearly see an increase of interest of HCI researchers and practitioners in the inclusion of gaze gestures which can greatly simplify and made more effective and efficient the communication between the human user and the computer. This is an example of a larger trend that the communication between the human and computer becomes more and more "physical", that is, using as much as possible, and as directly as possible, what can be read from how the human behaves.

Of course, such a gesture based interface can involve all what can be learned from movements of the human body, that is, to just name a few, from: face, hand, leg and foot, etc. to the whole body movement. Moreover, as the human involvement in the human–computer interaction can extend to the involvement of groups of agents, it

C. Biele, *Human Movements in Human-Computer Interaction (HCI)*,
Studies in Computational Intelligence 996,
https://doi.org/10.1007/978-3-030-90004-5_1

also makes sense to consider movement in the sense of various movements of groups of users of the HCI, even finally movements at the level of a society of HCI users.

These important, explicitly human centric issues in the development, design, analysis and implementation of the HCI systems are discussed in the book. A comprehensive state of the art is given and is complemented with some original own proposals. Emphasis is on various types of human movement in the context of the human–computer interaction. However, as opposed to main works in the field which discuss the broadly perceived human movement related aspects in the HCI from a more formal, mathematical modeling based, point of view, followed by much detail on the use of modern IT/ICT tools for their implementation, in this book the above mentioned analyses will be more explicitly based on relevant research results from psychology and psychophysiology. By necessity, some more relevant results obtained in these sciences are used showing how they can extend and improve the HCI development. However, The inclusion of these aspects in the HCI field is important because the human being is in this area the key element and we should take into account as much as possible what is known about the human behavior and how this can be transformed into algorithms to be employed in the HCI systems.

By following this path, hopefully, this book can be interesting and useful to a large community of users who are interested or involved in the development, design, analyses and implementation of various HCI systems. These communities can include both software and hardware specialist, psychologists, cognitive scientists, to just mention a few.

The human–computer interaction can be defined as a loop in which users communicate their intentions in a manner that is understandable to machines. A reaction occurs in response to this input, and users receive feedback communicated in a manner understandable to humans, to which they themselves can then react and issue further commands; thus, completing the loop. Although at first glance, the process appears static, motion forms an integral part of any type of human–computer interaction currently in use: from finger and hand movements during the use of keyboards and mice; to finger-operated controllers that connect to modern games consoles; to eye movement tracking systems designed for those who are unable to use their limbs; to full-body movement tracking used for interactions in virtual reality (VR).

Recently, the role of motion in human–computer interaction has played an increasingly important role; this is partly attributable to the development of mobile technologies. Users' locations in urban environments or how they move in spaces may act as signals not only for navigation systems, shared mobility car, or electric scooter rental systems, but also for systems that assign tasks on spatial crowdsourcing platforms.

The above examples demonstrate clearly that although motion has been an essential component of human–computer interaction since the dawn of computerization, it currently plays a key role in virtually all methods of computer interaction.

Although deemed to be of a marginal importance during the age of text-based interfaces, motion has grown to attract considerable attention, and has become an essential element of screen-based interfaces. Smartphone and computer screens display animated windows, progress bars, and spinning mouse pointers to indicate that systems are busy or are working in background—all of which influences how interactions look and how they are perceived by users. Robotic solutions are also becoming more popular; the best-known of them is likely the Roomba automated vacuum cleaner. Movement is their key functionality, and it influences users' experiences and opinions of such robots.

In order to be understood fully, motion must be studied in the context of psychology and neurophysiology. Movement is a basal capability that has shaped the human mind. Some even consider thinking to be an abstract form of movement, arguing that the two evolved in tandem as humans moved to hunt and source food.

As a consequence of such assertions, the theory of embodied cognition has gained in prominence since the 1990s. It posits that cognitive processes involve not only the separate, abstract mind— as traditional theories of cognition postulate—but the entire body. The traditional approach is somewhat narrow: it professes that cognitive processes occur independently from those—be it sensory processing or motor control—in the body.

Contrarily, the embodied cognition theory postulates that in order for cognitive processes to be understood, they must be viewed from a broader perspective; in other words, the complete range of perception and movement processes must be considered, as they are linked directly to cognition.

1.2 Book Composition

Chapters 2–6 are dedicated to computer interaction from the perspectives of specific body parts. The chapters discuss facial movements (in particular, those related to the expression of emotions); eye movements (control interaction with the use of eye-movements and measurement of cognitive loads); hand movements (use of keyboards and mice, and bioidentification based on movement patterns); leg movements (new methods of interaction involving the legs and feet); and whole-body movements (identification based on body movement patterns and forms of interaction that utilize new technologies). Chapters 7–10 illustrate concepts related to movement in virtual spaces and in the environment, its perception, and its relationship with cognitive processes in human–computer interaction. Chapter 11 focuses on the future of various types of interaction, including brain-computer interfaces and voice control—which, unlike the subjects discussed in the preceding chapters, do not rely on movement directly.

Brief summaries of each chapter are presented below. Chapter 2, "Face movement", focuses on face movement in the context of human–computer interaction. Although discussed primarily in the context of emotional expressions, face movement can also be analyzed independently from emotions. The first section outlines the brain processes responsible for emotions and generating facial expressions. It also offers broad information on surface electromyography (sEMG)—a popular method of recording the activity of facial muscles and face movement that is also used in human–computer interaction. Later sections explain the distinctions between two approaches to the recognition of facial muscle movements: one based on video signals, and the other on the measurement of muscle activity. The methods are analyzed with regard to their potential applications in the field of human–computer interaction.

Chapter 3, "Eye movement", studies concepts related to eye movement in human–computer interaction. It presents general information on the anatomy and physiology of the human eye, and outlines the key types of eye movement. It also presents methods of eye tracking and some of the fields in which they can be applied. The chapter discusses issues pertaining to cognitive load, and how its intensity in relation to human–computer interaction and hypertext reading can be determined using eye activity measurements. The final section is dedicated to the employment of eye-tracking technology as a method of interaction and as a source of signals in communication with computers.

Chapter 4, "Hand movement", analyzes the use of traditional computer peripherals that can only be operated manually, such as keyboard and mice. These relate closely to movement of the hands, fingers, and forearms. How users operate such devices allows considerably more information to be acquired than records of typed text, clicks, and mouse movements; it is possible, for instance, to identify users and to monitor their emotional states using peripherals. The first section outlines how keyboards are used to identify users and their emotions by application of keystroke dynamic analysis. It also explores differences in learning efficiency between keyboard use and traditional methods of recording information. The second section discusses the use of mice in identification of users and their emotional states.

Chapter 5, "Leg and foot movement", explores human–computer interaction involving the feet and legs. It presents the elements of non-verbal communication that affect foot position and movement, as well as how the latter is linked to aspects of human psychology. It also presents key information on the anatomy of the feet, and describes how that knowledge may be employed in interactions with computer systems. The second half of the chapter outlines technologies that utilize the feet as a means of interaction. It presents examples of direct and indirect interfaces, in addition to those based on sensors installed in the environment—such as sensing floors.

Chapter 6, "Whole-body movement", discusses whole-body movement in the context of human–computer interaction. It discusses the recognition and classification of body movements with the use of motion capture systems and video signal analysis. It also presents some of the practical applications of such systems, including automatic recognition of sign language and identification of individuals. The second half of the chapter reflects on the use of body movement as a method of interaction with computers and machines. New paradigms of interaction involving the whole body—including remote operation (for example, drone control)—are also discussed.

Chapter 7, "Movement in virtual reality", presents concepts related to movement in virtual reality (VR), placing emphasis on its prospective applications in psychological studies. Behavioral measurements are a key tool in such studies—which presently, are almost invariably conducted in unnatural and experimental settings, and entail considerable difficulty in obtaining measurements. The chapter illustrates examples of VR use in studies of personality and sex differences, in psychological and psychiatric practice—for example, as a supporting tool in addiction treatment; and in more general applications that utilize the technology in user movement tracking—such as in the learning of dance movement patterns. The final section discusses negative sensations experienced while using VR systems.

Chapter 8, "Movement in the environment", discusses users' movements within their environments—both indoors and over longer distances (in cities, for instance). It reflects on the technologies used to locate users and their potential applications, including those in security and in ecology. The chapter then presents issues relevant for human–computer interaction, including the impact of navigation technologies on human behavior, the protection of privacy in solutions that employ location monitoring (for example, shared mobility), and users' views on privacy concerns. The chapter concludes with an explanation of relatively new phenomena in user movement and human–computer interaction—including spatial crowdsourcing and shared mobility—in the context of user motivation and privacy.

Chapter 9, "Perception of movement", is concerned with the perception of movement that occurs during human–computer interaction. The phenomena discussed include, but are not limited to, perception of biological movement and deduced movement. The chapter also outlines the concept of mirror mechanisms. The later sections explain the potential effects of perceived motion on how the properties of interfaces are perceived—such as their attractiveness, or their perceived performance. The chapter culminates with a summary of recent research on how movement affects perceptions of avatars.

Chapter 10, "Movement, cognition, and learning", examines the relationship between movement and mental and cognitive function, as well as its potential application in new technologies. The first section presents the cerebral mechanisms responsible for associating physical with mental activity, and discusses examples of their influence on cognitive and emotional processes, as well as learning. The second section focuses on physical activity as an element of human–computer interaction. The final section presents the subject of physical activity and cognitive functions from the perspective of immersive VR technology. The potential of VR stems from its ideal compatibility with the study of motion and to its well-established relationship with mental functioning. The chapter also reflects on immersive VR as a potentially effective motivator in increasing individuals' physical activity, with the goal of improving their mental functioning.

Chapter 11, "Future directions", presents some answers to these questions: what does the future hold for motion-based interaction methods; will increasingly popular movement-based interaction concepts and inventions, such as immersive VR, hasten their already rapid development; or will they be supplanted by other solutions, such as interfaces capable of reading brain activity directly, or those that recognize voice commands? Such interfaces do not require movement, and each enables new perspectives—for users and technologists alike. The development of technologies used to record movement, such as immersive VR, is also of high importance for other branches of science, including psychology. To date, no solution has facilitated

Chapter 2
Face Movement

2.1 Introduction

Face perception is one of the most highly developed visual abilities that humans possess. From birth, babies prefer looking at faces than other objects, and tend to fix their eyes on faces for extended periods of time. Despite the enormous number of faces humans see and are capable of remembering, we generally have little difficulty recognizing others—even those whom we have not seen for many years. Perception of faces is not limited, however, to the identification of individuals, as they are most often perceived in the social context—emotional expressions also offer information that is crucial for communication. In the case of human–computer interaction, the face can serve as a source of data to read users' emotional states or to control interactions. The most common methods of recording the activity of facial muscles are based either on registration and analysis of the electrical activity of facial muscles, or on recording video signals of faces and on computer detection of movement of individual parts of faces.

2.2 Perception of Emotional Expressions

Face perception is one of the most highly developed visual abilities in humans. Infants prefer looking at faces than other objects—even when those objects resemble human faces. It is known that the face is a rich source of information that allows individuals' identities, ages, genders, and ethnicities to be discerned. This information is accessible due to the perception of constant traits, such as the shape of the eyes, nose, and mouth. Above all, the face plays an important role in interpersonal communication [1], which is partially enabled by the expression of emotions on human faces. These have been studied since the nineteenth century, when Charles Darwin first explored the subject in his famous work, *The Expression of Emotions in Man and Animals.*

C. Biele, *Human Movements in Human-Computer Interaction (HCI)*,
Studies in Computational Intelligence 996,
https://doi.org/10.1007/978-3-030-90004-5_2

Darwin asserted that emotional expressions revealed individuals' emotional states. In the 1970s, Ekman demonstrated that emotional expressions were universal. Notion worth mentioning is that some researchers in the area on the expression of emotions claim that emotions are not merely expressions, but 'signals of intent [2, 3].

2.3 Neurophysiological Bases of Emotional Expression Recognition

Since emotions and emotional expressions are universal, they are believed to originate in the brain. In order to answer the question whether particular emotions are neural-based, different brain imaging methods have been employed, including functional magnetic resonance imaging (fMRI), positron emission tomography (PET), and single-photon emission computed tomography (SPECT). These methods measure blood flow through the brain and metabolisms—particularly glucose compounds—in individual brain structures. It is assumed that greater activity in a given brain area increases blood flow, oxygen concentration, and glucose consumption. These methods allow researchers to observe the activity of individual brain structures, including those responsible for the perception of emotional expressions.

In this type of study, usually healthy subjects are asked to look persistently at or recognize the emotions of faces in photographs [4]. In addition to typical face photos, computer-processed photos are sometimes used that differ in their intensity of a given emotion [5]. These types of study chiefly aim to understand the neural correlates of negative emotions, including fear.

2.3.1 Fear

Fear is associated with the amygdalae—each of which forms a part of the limbic system, and consists of a number of nuclei that differ morphologically, biochemically, and functionally. The amygdalae are characterized by the large number of internal and external connections through which information from all cortical sensory areas is transferred to them. Research conducted by LeDoux et al. shows that the amygdalae also maintain direct connections with the thalamus, which ensures faster, shorter paths (bypassing the cortex) between them. [6, 7]. LeDoux claims that such connections enable rapid reactions to even the faintest signals of impending danger. The assumption that the amygdalae play a key role in fear is based both on animal studies and on the clinical trials of human subjects. The results of neuroimaging studies confirm the importance of the amygdalae in the processing of information on fear [8]. It has been demonstrated that the observation of faces expressing fear stimulates the amygdalae, and in particular, the left one [9, 10]. Additionally, many clinical studies indicate that both patients with unilateral and bilateral amygdala damage are unable to recognize the emotion of fear in faces presented to them [11, 12]. Interestingly, the more intense the emotions are, the more the structure is stimulated [5]. Brain activity measurements have shown that the observation of faces expressing fear also activates cortical structures: the medial prefrontal cortex of the right hemisphere, the dorsolateral part of the left frontal cortex, the cingulate gyrus, the entire temporal gyrus, and the right fusiform gyrus [13]. The activity of the last region seems unambiguous: the fusiform gyrus—which is also known as the "face area"—is stimulated when a face is perceived. The activation of the frontal areas can be explained by the existence of a neuronal loop between the left amygdala, the left ventrolateral part, and the right dorsolateral part of the frontal cortex that participates in the processing of negative emotions [14].

2.3.2 *Anger and Sadness*

Anger and sadness receive much attention owing to their social importance. Correctly recognizing them requires the ability to understand specific situations, feelings, and others' motivations. It has been shown that people who demonstrate signs of psychopathy, which is characterized by emotional depletion, are unable to correctly recognize anger nor sadness in the faces of others [15]. Great importance can be attributed to the frontal cortex—particularly the orbitofrontal cortex—in this regard. Evolutionarily, it is the most recently developed area of the brain. It is responsible for the ability to predict and for abstract thinking. It also regulates a number of social behaviors [16, 17]. Neuroimaging studies have confirmed that the medial prefrontal cortex is involved [4, 18] in the processing of information on anger and sadness [19]. Observations of expressions of sadness also enhance activity in the temporal areas [20]—sometimes only on the right side, and in the cingulate cortex [21].

Activation of the orbitofrontal cortex accompanies recognition of both anger and sadness [13, 21]. The cingulate cortex—or, more specifically, the anterior [21] and posterior parts of the right cingulate gyrus and the middle part of the left temporal gyrus—is also stimulated. Research results indicate that the recognition of both sadness and anger is associated with the activity of the orbitofrontal cortex and of the cingulate cortex, which is involved in various psychological functions. It has been demonstrated that damage to the anterior cingulate cortex can lead to movement disorders and disruptions in the cognitive processes related to emotions, memory, and attention. It appears that attention plays a significant role in identifying anger, as anger is an emotion of a social nature, and can alert to approaching danger; its presence demands that individuals focus all their attention on the threatening stimulus. This might explain why the cingulate gyrus is stimulated when identifying angry faces. The joint activation of the cingulate cortex during the recognition of sadness and anger suggests that the neuronal systems responsible for recognizing these emotions overlap to some extent.

2.3.3 Disgust

Research shows that observing faces expressing disgust activates the medial prefrontal cortex [8] and the insula [9, 13]. The activity of the insula and the surrounding cortex—which is also known as the primary gustatory cortex—has been demonstrated in studies on unpleasant tastes and odors [22]. The perception of disgust also activates the subcortical structures: the right putamen, the globus pallidus of the thalamus, and, most interestingly, the right caudate nucleus. Patients with Huntington's disease, which is caused by degeneration of the caudate nucleus, exhibit specific difficulties pertaining to the expression of disgust.

2.3.4 Happiness

Studies on positive emotions are essentially limited to happiness. Smiling seems to be an innate ability; it is known that even blind and deaf infants are able to smile shortly after birth. Interestingly, no disorders in recognizing the expression of happiness have been identified in any group of patients with brain damage [23]. Analysis of brain activity during the experience of happiness shows involvement of the medial prefrontal cortex [18, 24, 25], the right temporal area, and the left parietal lobe [5]; the thalamus, the right putamen, and the amygdalae [26] also displayed excitement. In other reports, no significant increase in brain activity has been found when experiencing happiness; on the contrary, there was a decrease in excitement of the right prefrontal area and on both sides of the temporoparietal area. Alternating increases and decreases in activity have been observed in the anterior and posterior parts of the cingulate cortex. These results suggest that there is no specific pattern for brain activity in the case of recognition of happiness. It is noteworthy, however, that the amygdalae, which are specifically related to fear, activate when an individual perceives happiness.

2.3.5 Surprise

There are few formally published neuroimaging studies on surprise. It is the shortest emotional expression, lasting less than a second. It also appears to be the most dynamically expressed one, and is often referred to as a "transitory" emotion [27], which, depending on the situational context, evolves into happiness or fear. It can therefore be expected that the perception of surprise involves the amygdalae. This can also be inferred, albeit indirectly, from studies on individuals with damaged amygdalae, which show that, in addition to disruptions in recognizing fear, the subjects also exhibited problems in recognizing surprise [12].

2.3.6 Summary

The cerebral activity measured in the process of recognizing happiness, sadness, anger, fear, and disgust indicate the existence of a specific system of cortico-subcortical stimuli. The structure activated is almost invariably the medial prefrontal cortex, which might constitute a common neuronal basis for all emotions experienced by humans. This activation occurs both in social emotions, such as happiness, sadness, or anger; and in biological emotions, such as fear or disgust. The involvement of the prefrontal cortex in such a variety of emotions can be attributed to its specific nature and to that of its connections. The anterior part of the cortex (the orbitofrontal cortex) has particularly strong connections with the cingulate cortex, the temporal area [28], the basal nuclei [29, 30], and the amygdalae [31]—in other words, with the part of the limbic system associated with primal emotions, such as fear. This is likely why simultaneous activation of the prefrontal cortex and the amygdalae is so frequently observed. Some researchers even claim that only the frontal part of the prefrontal cortex (the orbitofrontal cortex) is involved in emotional processes. This approach asserts that the orbitofrontal cortex is involved in the development of various types of emotion—it is the foundation of the common neural system for all emotions [32, 33].

2.4 Face Movement During Emotional Expression

Research on the activity of facial muscles during affective reactions was initiated in the mid-1970s by Schwartz in his series of studies on imagery [34, 35]. He discovered that visualizing positive or negative situations or things influenced the activity of both the major zygomatic muscle and the corrugator supercilii muscle: positive images increased the activity of the zygomatic muscle, while negative ones amplified that of the corrugator supercilii. Electromyography studies have also demonstrated different affective reactions during the perception of various types of stimulus, such as unpleasant sounds [36], affectively evocative photographs [37], or emotional words [38]. Extensive and comprehensive EMG research focuses on the perception of emotional expressions and the accompanying muscle responses [39–42]. Studies demonstrate that expression occurs even faster in these cases than it does when presented with threatening stimuli. The activity of the major zygomatic muscle, which is responsible for raising the corners of the mouth when smiling, differed significantly during the presentation of happy and angry faces 300 ms after seeing the stimulus; in the case of the corrugator supercilii muscle, which is responsible for lowering the eyebrows, the difference occurred after 500 ms. It is important to note that these differences were noticeable even during the initial stimulus presentations. The results of these experiments demonstrate that muscle responses render it possible to distinguish emotional responses within a few hundred milliseconds of presentation. Reactions during the perception of emotional expressions occur rapidly, and have also been observed in young children [43]. The presence of such swift and inborn responses support the hypothesis that emotional responses to faces can occur relatively quickly and unconsciously.

Similar reactions are triggered when a photo of a face is displayed for a very short time (30 ms)—so short that the subjects are unable to consciously perceive it [44]. This indicates that facial reactions produced in response to the perception of emotional expressions occur automatically. Interestingly, the activity of the appropriate facial muscles is also recorded when subjects are requested not to respond to the perceived stimuli [41], which proves that it is impossible to exercise full conscious control over emotional expression. These effects are not limited to expressions of anger and happiness; studies on the perception of sadness, anger, fear, surprise, disgust, and happiness show increased activity of the corrugator supercilii, the major zygomatic, the lateral frontalis, the depressor supercilii, and the levator labii superioris muscles [45]. These results support the hypothesis that the automatic affective response system is one of the most fundamental and that the responses of the facial muscles are controlled by fast-acting and biologically-based "affective programs" [46].

2.5 Facial Expressions in Human–Computer Interaction

The most common theory of the universality of emotional expressions developed by Ekman, [47], postulates that seven fundamental expressions of emotion exist—anger, contempt, disgust, fear, joy, sadness, and surprise. Ekman's research on emotional expressions also developed an approach to their recognition, and his findings continue to be applied in the area of human–computer interaction. Ekman characterized the movements of individual parts of the face, which he termed action units. The Facial Action Coding System describes how the movement of individual action units translates into specific universal emotional expressions [48, 49]. This can be illustrated by the expression of happiness which, according to the system, is a combination of AU-6 (cheek raiser) and AU-12 (lip corner puller). The system comprises 46 action units in total, which cover every area of the face. During human–computer interaction, changes in the areas of individual action units can be recognized with the use of various methods, such as muscle activity recording and computer algorithms. Solutions based on recognizing emotional expressions are applied in many areas of life, including road safety and healthcare. The task itself, however, is challenging to implement due to the ambiguous and dynamic nature of emotional expression.

2.5.1 Video-Based Facial Expression Recognition

Computer emotional expression recognition methods utilize computational algorithms to recognize expressions in images or in video streams that display human faces. They are usually limited to the so-called basic expressions: anger, fear, disgust, surprise, sadness, and happiness, and are based on the Facial Action Coding System proposed by Ekman and Friesen [48]. Typically, emotion recognition systems perform this task in two key stages: they first extract parts of a face from an image, before proceeding to classification. The details and technicalities of the approaches differ considerably, both in the extraction and the classification stages. In practice, however, they almost always rely on action units or combinations thereof.

Feature extraction solutions can be divided into two primary groups: those that analyze faces as a whole, and those that focus solely on specific parts or areas of faces. In the former case, three types of method are commonly used—motion extraction based on optical-flow [50], difference images [51], or methods based on single images [52]. Optical flow can be defined as the movement of objects between consecutive video frames, which is operationalized as the distribution of the speed of moving brightness patterns in images. When tracking the movement of parts of the face with the use of optical flow, so-called "dense" optical flow is employed, which considers all pixels in the images. This method can be rendered ineffective if the brightness of the image changes, or if the image is poorly equalized. Parts of the face that express emotions can also be detected by analyzing the differences between the examined face and a neutral face. One advantage of this approach is that it is insensitive to changes in lighting. Other methods involve analyzing static images, which allows the possibility to detect changes in the position of parts of the face, as well as in the appearance of the surface of the skin. The weakness of these methods, however, is that they fail to account for the dynamics of expression and its evolution over time, both of which form key elements of emotional expression [53].

The second feature extraction approach focuses on analysis of parts of the face, rather than its entirety. Although this approach limits the ability to recognize the full range of expression, it also demands less computing power. As every action unit is involved in the formation of basic expressions to a different extent, local methods focus on those areas that are most relevant to the formation of basic expressions. These areas include the eyes, eyebrows, and mouth. Movement of parts of the face and changes in its shape are recognized by tracking sets of points around the most relevant parts of the face. The points are then normalized with the use of a reference face [50].

The recognition of action units is the initial step in the process of computer-based expression recognition; the next is to classify the activities recognized, which can be either spatial or spatiotemporal. In the case of spatial classification, neural networks are used, which can receive a variety of inputs, such as PCA coefficients or Gabor-wavelet coefficients [51, 54]. Methods commonly used include Support Vector Machines and Linear Discriminant Analysis [52]. One major limitation of this approach is that it fails to consider the temporal characteristics of emotional expressions, which are dynamic by nature [55]; another is that networks can only be trained on posed and controlled expressions of emotion, as too many action units are activated simultaneously in the case of spontaneous emotions. Expressions are ambiguous by nature and are able to shift dynamically. This cannot be detected by systems that use static prototypical emotional expressions, however there are some solutions based on the spatiotemporal signal [56, 57].

Despite many years of research on computer-based recognition of emotional expressions of the face, the results achieved by such systems remain unsatisfactory. New approaches based on empirical mode decomposition [58] or fuzzy logic [59] have been proposed to improve their efficiency. Solutions that allow for the simultaneous recognition of the expressions of multiple users also exist [60].

2.5.2 Applications of Computer-Based Face Movement Recognition

Facial movements recognized by computers can be used as signals to control external devices. Viet et al. offered a solution designed to control wheelchairs [61]. In their application, head movements are used to turn wheelchairs, while the shape of users' mouths (open/closed) is used to control their movement (forward/stop). Faces were detected using Haar-like features, while lip movements were classified using Principle Component Analysis.

Computer methods can also be employed to detect pain based on the movement of facial muscles. This is possible because, as in the case of emotions, the activity patterns of individual faces are broadly similar in the expression of pain [62, 63]. Studies suggest the existence of one [64] or four [63] characteristic activity patterns involving the eyebrows, nose, and mouth. For that reason, action unit recognition is the most common approach. Using support vector machines and working on input data with information on the activity levels of twenty action units, Littlewort et al. were able to differentiate feigned pain expressions from genuine ones with 72% accuracy [65]. The effectiveness of such studies indicates that the methods of recognizing emotional expressions currently being developed could potentially be used for research on other psychological phenomena. It is important to note, however, that although the efficiency of programming methods is satisfactory in the case of recognizing prototype and intensive expressions, humans continue to outperform machine software in the case of less intense or not entirely stereotypical expressions [66].

2.5.3 Surface Electromyography for Movement Detection

One method used in psychophysiological studies involves surface electromyography (EMG), which measures the electrical activity of striated muscle tissue (skeletal muscles). When a body part moves, impulses from the neurons that innervate muscle fibers (motor neurons) stimulate muscle cells (muscle contractions), and change their action potential. These changes can be recorded using EMG [67]. Dutch biologist Jan Van Swammerdam (who is reportedly portrayed in Rembrandt's masterpiece,

The Anatomy Lesson of Dr. Nicolaes Tulp), and Italian Francesco Redi were the pioneers in research on the electrical activity of muscles in the eighteenth century. During his experiments on frogs, Van Swammerdam discovered that the mechanical stimulation of nerves caused muscles to contract. His research also reputedly involved the electrical stimulation of muscles (almost a century and a half before Luigi Galvani, who is considered to be the father of neurophysiology!). Redi was the first to notice that the movement of muscles was related to electrical activity. He also studied electric rays (*Torpedo torpedo*) and discovered that the fish possessed specialized muscle tissue responsible for generating electrical current, which they use to stun the small fish and crustaceans on which they feed. Advancement of the research into the electrical activity of muscles came with the invention of the electric battery by Alessandro Volta. In the late eighteenth century, Luigi Galvani demonstrated that the electrical stimulation of muscles caused them to contract.

EMG is used in many fields of psychology, and is of particular interest to researchers who study mental processes, motivation, and emotions. Muscles hold a special place in psychology because they allow humans to interact with their environment. Some of the effects of muscle activity, such as changes in posture, are easily observable, and others less so; nevertheless, they all allow the inference of indirect conclusions on the processes in which researchers are interested. The first efforts to explore this area came in the nineteenth century, when attempts were made to demonstrate that muscle activity was connected with mental phenomena. In the following years, scientists attempted to explain how the solving of mental tasks related to muscle tone.

One intriguing phenomenon studied with EMG is so-called "subvocal" speech –increases in the activity of facial muscles involved in speech production during the performance of various cognitive activities, such as, reading, problem-solving, and recollecting. The increase in activity is larger in the case of poor readers, or in the case of difficult reading material [68].

The research conducted by Jacobson in the 1930s is equally compelling [69]. His subjects were tasked with imagining various activities, such as reciting poems, lifting heavy things, or viewing objects. Jacobson discovered that the muscle tone around the mouths of the subjects increased when they were asked to visualize or recall linguistic activities, and that their eyes moved up and down when they were prompted to visualize the Eiffel Tower.

Surface EMG is also used in studies on human–computer interaction, in which the activity of facial muscles is sometimes used as the source of signals. Measuring users' emotional states can prove a valuable assessment element in human–computer interaction. Affect information can provide details indirectly pertaining to the attractiveness of interfaces. Much of the human experience in human–computer interaction, which results from emotional responses, can be captured using EMG. According to the circumplex model of affect [70], all basic emotions can be placed into one of two dimensions—valence and arousal. Happiness and anger, for instance, are similar in level of arousal, but differ fundamentally in the dimension of that arousal. Facial surface EMG, which measures the electrical activity of the facial muscles, can be used to assess positive and negative emotional valence; the specific systems of muscle activity for individual emotional expressions, which are related to users' emotions, enable this.

2.5.4 *Facial Muscle Activity as Signals in HCI*

Facial movements can be used as input signals in interaction with computers and other devices. In this case, the most frequently used signals are the surface electromyographic signal (sEMG) and the electrooculogram (EOG). The most promising implementations of these technologies lies in solutions for those suffering from health conditions that prevent them from effortlessly moving their hands or speaking—the activity of facial and eye muscles could become an effective communication channel, allowing users to control wheelchairs, for example [71]. Such solutions rely on signals such as conscious eye blinking or lips tightening [72]. Systems based on similar principles are being designed specifically for human–computer interaction. The activity of the eye muscles is used to control cursor movement on screens, while voluntary eye blinks control mouse clicks [73]. There are also solutions for input devices, such as keyboards controlled by eye and facial muscles [74].

Muscle signals are usually assigned controlling commands to be used in a single channel (for instance, the activity of one facial muscle) to control a single action of a system. The signals are often paired, but are nevertheless processed separately. In the studies by Williams and Kirsch [75], cursor movements along a vertical axis were controlled by the EMG signal of two facial muscles—the frontalis muscle and the mentalis muscle. The cursor moved higher when the electrical activity of the frontalis muscle was greater than that of the mentalis muscle, and down in the opposite case. This naturally limits the number of potential commands available in such systems, as the number of facial muscles that can be used is finite. Moreover, if different muscles are used, the signals of each must be initially processed and calibrated separately, as individual muscles have different levels of electrical activity—due, for example, to differences in size.

The solution is an approach that recognizes multi-muscle activation patterns, which assumes that different facial expressions have different, but repetitive movement patterns. This means that pattern recognition algorithms can be used to process the electrical activity signals coming from different muscles. The key advantage of the approach over "one muscle, one action" systems lies in its ability to perform a high number of actions while recording the activity of a low number of muscles. It has been used, for instance, in muscle interfaces that read hand gestures to control exoskeletons [76–78]. Lu and Zhou [79] utilized the same approach when developing the Facial Movement Machine Interface (FMMI). The architecture of the system is linear in character. In the first stage, facial muscle movements are recorded by EMG sensors. The signals received are subsequently recognized by a pattern recognition system, which sends the resulting patterns of muscle activity to a module that transforms tem into commands to control a cursor—ultimately enabling human–computer interaction. The system recognizes five movement patterns (plus no movement) using only four electrodes placed on the risorius, mentalis, and temporalis muscles. This allows users to move cursors in four directions and to perform two types of click—long ones and short ones. The accuracy of the pattern recognition was approximately 96%. Tests with users proved the method effective for typing on virtual keyboards and for drawing. Pattern recognition systems can also use combined EMG and EEA signals, which yields better results than processing the signals independently [74].

2.5.5 Facial Muscle Activity as Emotion Signals

Facial muscles are used not only as signals to control computers, but also as a source of information on users' emotions at any moment, and can be applied to research on emotions when using a computer. Additionally, they can serve as elements of affective computing systems, in which users' emotions are scanned; thus, enabling the systems to run more efficiently.

Psychophysiological methods of measuring the activity of facial muscles and, consequently, users' emotional states, are also used to measure the quality of user experience—which frequently conflicts with the parameters strictly related to functionality, such as the time required to complete tasks. The emotional component can influence the perceived attractiveness and quality of interactions, and this, in turn, influences the overall perception of computer systems [80]. Research on the attractiveness of video games in which emotional indicators obtained from facial movements were used has been conducted by Mandryk et al. [81]. In addition to measuring facial muscle activity with surface EMG, the authors utilized other psychophysiological measures, such as electrodermal activity, and breathing and heart rate parameters. Solutions for determining emotional states based on fuzzy logic have also been proposed [82]. Using several psychophysiological signals (including the activity of facial muscles), researchers are able to determine the attractiveness of interactions and the degree of users' involvement. In addition, affective computing solutions are being developed for mobile devices [83, 84].

References

1. Mehrabian, A.: Silent Messages: Implicit Communication of Emotions and Attitudes. Wadsworth Publishing Company (1981).
2. Chovil, N.: Social determinants of facial displays. J. Nonverbal Behav. **15**, 141–154 (1991)
3. Jakobs, E., Manstead, A.S.R., Fischer, A.H.: Social motives and emotional feelings as determinants of facial displays: the case of smiling. Pers. Soc. Psychol. Bull. **25**, 424–435 (1999)
4. George, M.S., Ketter, T.A., Parekh, P.I., Horwitz, B., Herscovitch, P., Post, R.M.: Brain activity during transient sadness and happiness in healthy women. Am. J. Psychiatry. **152**, 341–351 (1995)

5. Morris, J.S., Frith, C.D., Perrett, D.I., Rowland, D., Young, A.W., Calder, A.J., Dolan, R.J.: A differential neural response in the human amygdala to fearful and happy facial expressions. Nature **383**, 812–815 (1996)
6. Ledoux, J.: The Emotional Brain: The Mysterious Underpinnings of Emotional Life. Simon and Schuster (1998)
7. Ghashghaei, H.T., Hilgetag, C.C., Barbas, H.: Sequence of information processing for emotions based on the anatomic dialogue between prefrontal cortex and amygdala. Neuroimage **34**, 905–923 (2007)
8. Phillips, M.L., Young, A.W., Senior, C., Brammer, M., Andrew, C., Calder, A.J., Bullmore, E.T., Perrett, D.I., Rowland, D., Williams, S.C., Gray, J.A., David, A.S.: A specific neural substrate for perceiving facial expressions of disgust. Nature **389**, 495–498 (1997)
9. Phillips, M.L., Young, A.W., Scott, S.K., Calder, A.J., Andrew, C., Giampietro, V., Williams, S.C., Bullmore, E.T., Brammer, M., Gray, J.A.: Neural responses to facial and vocal expressions of fear and disgust. Proc. Biol. Sci. **265**, 1809–1817 (1998)
10. Breiter, H.C., Etcoff, N.L., Whalen, P.J., Kennedy, W.A., Rauch, S.L., Buckner, R.L., Strauss, M.M., Hyman, S.E., Rosen, B.R.: Response and habituation of the human amygdala during visual processing of facial expression. Neuron **17**, 875–887 (1996)
11. Calder, A.J., Keane, J., Manes, F., Antoun, N., Young, A.W.: Impaired recognition and experience of disgust following brain injury. Nat. Neurosci. **3**, 1077–1078 (2000)
12. Adolphs, R., Tranel, D., Hamann, S., Young, A.W., Calder, A.J., Phelps, E.A., Anderson, A., Lee, G.P., Damasio, A.R.: Recognition of facial emotion in nine individuals with bilateral amygdala damage. Neuropsychologia **37**, 1111–1117 (1999)
13. Sprengelmeyer, R., Rausch, M., Eysel, U.T., Przuntek, H.: Neural structures associated with recognition of facial expressions of basic emotions. Proc. Biol. Sci. **265**, 1927–1931 (1998)
14. Iidaka, T., Omori, M., Murata, T., Kosaka, H., Yonekura, Y., Okada, T., Sadato, N.: Neural interaction of the amygdala with the prefrontal and temporal cortices in the processing of facial expressions as revealed by fMRI. J. Cogn. Neurosci. **13**, 1035–1047 (2001)
15. Patrick, C.J., Bradley, M.M., Lang, P.J.: Emotion in the criminal psychopath: startle reflex modulation. J. Abnorm. Psychol. **102**, 82–92 (1993)
16. Sleezer, B.J., LoConte, G.A., Castagno, M.D., Hayden, B.Y.: Neuronal responses support a role for orbitofrontal cortex in cognitive set reconfiguration. Eur. J. Neurosci. **45**, 940–951 (2017)
17. Torregrossa, M.M., Quinn, J.J., Taylor, J.R.: Impulsivity, compulsivity, and habit: the role of orbitofrontal cortex revisited. Biol. Psychiatry. **63**, 253–255 (2008)
18. Lane, R.D., Reiman, E.M., Ahern, G.L., Schwartz, G.E., Davidson, R.J.: Neuroanatomical correlates of happiness, sadness, and disgust. Am. J. Psychiatry. **154**, 926–933 (1997)
19. Robinson, J.L., Demaree, H.A.: Experiencing and regulating sadness: physiological and cognitive effects (2009). https://doi.org/10.1016/j.bandc.2008.06.007
20. Paradiso, S., Robinson, R.G., Boles Ponto, L.L., Watkins, G.L., Hichwa, R.D.: Regional cerebral blood flow changes during visually induced subjective sadness in healthy elderly persons. J. Neuropsychiatry Clin. Neurosci. **15**, 35–44 (2003)
21. Blair, R.J., Morris, J.S., Frith, C.D., Perrett, D.I., Dolan, R.J.: Dissociable neural responses to facial expressions of sadness and anger. Brain **122**(Pt 5), 883–893 (1999)
22. Rolls, E.T.: A Theory of Emotion, and Its Application to Understanding the Neural Basis of Emotion (1986). https://doi.org/10.1159/000413540
23. Damasio, H., Bechara, A., Tranel, D., Damasio, A.R.: Double dissociation of emotional conditioning and emotional imagery relative to the amygdala and right somatosensory cortex. Soc Neurosci Abstr. 1318 (1997)
24. Damasio, A.R., Grabowski, T.J., Bechara, A., Damasio, H., Ponto, L.L., Parvizi, J., Hichwa, R.D.: Subcortical and cortical brain activity during the feeling of self-generated emotions. Nat. Neurosci. **3**, 1049–1056 (2000)
25. O'Doherty, J., Winston, J., Critchley, H., Perrett, D., Burt, D.M., Dolan, R.J.: Beauty in a smile: the role of medial orbitofrontal cortex in facial attractiveness. Neuropsychologia **41**, 147–155 (2003)

26. Hamann, S.B., Stefanacci, L., Squire, L.R., Adolphs, R., Tranel, D., Damasio, H., Damasio, A.: Recognizing facial emotion. Nature **379**, 497 (1996)
27. Posamentier, M.T., Abdi, H.: Processing faces and facial expressions. Neuropsychol. Rev. **13**, 113–143 (2003)
28. Rolls, E.T.: Memory systems in the brain. Annu. Rev. Psychol. **51**, 599–630 (2000)
29. Bush, G., Luu, P., Posner, M.I.: Cognitive and emotional influences in anterior cingulate cortex. Trends Cogn. Sci. **4**, 215–222 (2000)
30. Allman, J.M., Hakeem, A., Erwin, J.M., Nimchinsky, E., Hof, P.: The anterior cingulate cortex. The evolution of an interface between emotion and cognition. Ann. N. Y. Acad. Sci. **935**, 107–117 (2001)
31. Barbas, H.: Connections underlying the synthesis of cognition, memory, and emotion in primate prefrontal cortices. Brain Res. Bull. **52**, 319–330 (2000)
32. Hornak, J., Rolls, E.T., Wade, D.: Face and voice expression identification in patients with emotional and behavioural changes following ventral frontal lobe damage. Neuropsychologia **34**, 247–261 (1996)
33. Rolls, E.T.: The orbitofrontal cortex and emotion (2019). https://doi.org/10.1093/oso/978019 8845997.003.0006
34. Schwartz, G.E., Fair, P.L., Salt, P., Mandel, M.R., Klerman, G.L.: Facial muscle patterning to affective imagery in depressed and nondepressed subjects. Science **192**, 489–491 (1976)
35. Brown, S.L., Schwartz, G.E.: Relationships between facial electromyography and subjective experience during affective imagery. Biol. Psychol. **11**, 49–62 (1980)
36. Dimberg, U.: Perceived unpleasantness and facial reactions to auditory stimuli. Scand. J. Psychol. **31**, 70–75 (1990)
37. Cacioppo, J.T., Petty, R.E., Losch, M.E., Kim, H.S.: Electromyographic activity over facial muscle regions can differentiate the valence and intensity of affective reactions. J. Pers. Soc. Psychol. **50**, 260–268 (1986)
38. Hietanen, J.K., Surakka, V., Linnankoski, I.: Facial electromyographic responses to vocal affect expressions. Psychophysiology **35**, 530–536 (1998)
39. Dimberg, U., Thunberg, M.: Rapid facial reactions to emotional facial expressions. Scand. J. Psychol. **39**, 39–45 (1998)
40. Dimberg, U., Petterson, M.: Facial reactions to happy and angry facial expressions: evidence for right hemisphere dominance. Psychophysiology **37**, 693–696 (2000)
41. Dimberg, U., Thunberg, M., Grunedal, S.: Facial reactions to emotional stimuli: Automatically controlled emotional responses. Cogn. Emot. **16**, 449–471 (2002)
42. Dimberg, U., Hansson, G.O., Thunberg, M.: Fear of snakes and facial reactions: a case of rapid emotional responding. Scand. J. Psychol. **39**, 75–80 (1998)
43. Geangu, E., Quadrelli, E., Conte, S., Croci, E., Turati, C.: Three-year-olds' rapid facial electromyographic responses to emotional facial expressions and body postures. J. Exp. Child Psychol. **144**, 1–14 (2016)
44. Dimberg, U., Thunberg, M., Elmehed, K.: Unconscious facial reactions to emotional facial expressions. Psychol. Sci. **11**, 86–89 (2000)
45. Lundqvist, L.O.: Facial EMG reactions to facial expressions: a case of facial emotional contagion? Scand. J. Psychol. **36**, 130–141 (1995)
46. Tomkins, S.S.: Affect Imagery Consciousness: Volume I: The Positive Affects. Springer Publishing Company (1962)
47. Ekman, P.: Expression and the nature of emotion. Approach. Emot. (1984)
48. Ekman, P., Friesen, W.V.: Facial action coding system (2019). https://doi.org/10.1037/t27 734-000
49. Ekman, R.: What the Face Reveals: Basic and Applied Studies of Spontaneous Expression Using the Facial Action Coding System (FACS). Oxford University Press (1997)
50. Lien, J.J.-J., Kanade, T., Cohn, J.F., Li, C.-C.: Detection, tracking, and classification of action units in facial expression. Rob. Auton. Syst. **31**, 131–146 (2000)
51. Bazzo, J.J., Lamar, M.V.: Recognizing facial actions using Gabor wavelets with neutral face average difference. In: 6th IEEE International Conference on Automatic Face and Gesture Recognition, 2004. Proceedings, pp. 505–510 (2004)

52. Bartlett, M.S., Littlewort, G., Frank, M., Lainscsek, C., Fasel, I., Movellan, J.: Recognizing facial expression: machine learning and application to spontaneous behavior. In: 2005 IEEE Computer Society Conference on Computer Vision and Pattern Recognition (CVPR'05), vol. 2, pp. 568–573 (2005)
53. Rymarczyk, K., Biele, C., Grabowska, A., Majczynski, H.: EMG activity in response to static and dynamic facial expressions. Int. J. Psychophysiol. **79**, 330–333 (2011)
54. Tian, Y., Kanade, T., Cohn, J.F.: Evaluation of Gabor-wavelet-based facial action unit recognition in image sequences of increasing complexity. In: Proceedings of 5th IEEE International Conference on Automatic Face Gesture Recognition, pp. 229–234 (2002).
55. Scherer, K.R., Ellgring, H., Dieckmann, A., Unfried, M., Mortillaro, M.: Dynamic facial expression of emotion and observer inference. Front. Psychol. **10**, 508 (2019)
56. Zhang, Y., Ji, Q.: Active and dynamic information fusion for facial expression understanding from image sequences. IEEE Trans. Pattern Anal. Mach. Intell. **27**, 699–714 (2005)
57. Gu, H., Ji, Q.: Facial event classification with task oriented dynamic Bayesian network. In: Proceedings of the 2004 IEEE Computer Society Conference on Computer Vision and Pattern Recognition, 2004, CVPR 2004, pp. II–II (2004)
58. Ali, H., Hariharan, M., Yaacob, S., Adom, A.H.: Facial emotion recognition using empirical mode decomposition. Expert Syst. Appl. **42**, 1261–1277 (2015)
59. Bahreini, K., van der Vegt, W., Westera, W.: A fuzzy logic approach to reliable real-time recognition of facial emotions. Multimed. Tools Appl. **78**, 18943–18966 (2019)
60. McDuff, D., Mahmoud, A., Mavadati, M., Amr, M., Turcot, J., Kaliouby, R.E.: AFFDEX SDK: a cross-platform real-time multi-face expression recognition toolkit. In: Proceedings of the 2016 CHI Conference Extended Abstracts on Human Factors in Computing Systems, pp. 3723–3726. Association for Computing Machinery, New York, NY, USA (2016)
61. Viet, N.B., Hai, N.T., Van Thuyen, N.: Hands-free control of an electric wheelchair using face behaviors (2017). https://doi.org/10.1109/icsse.2017.8030831
62. Prkachin, K.M., Solomon, P.E.: The structure, reliability and validity of pain expression: evidence from patients with shoulder pain. Pain **139**, 267–274 (2008)
63. Kunz, M., Lautenbacher, S.: The faces of pain: a cluster analysis of individual differences in facial activity patterns of pain (2014). https://doi.org/10.1002/j.1532-2149.2013.00421.x
64. Prkachin, K.M.: The consistency of facial expressions of pain: a comparison across modalities. Pain **51**, 297–306 (1992)
65. Littlewort, G.C., Bartlett, M.S., Lee, K.: Faces of pain: automated measurement of spontaneousallfacial expressions of genuine and posed pain. In: Proceedings of the 9th international conference on Multimodal Interfaces, pp. 15–21. Association for Computing Machinery, New York, NY, USA (2007)
66. Yitzhak, N., Giladi, N., Gurevich, T., Messinger, D.S., Prince, E.B., Martin, K., Aviezer, H.: Gently does it: humans outperform a software classifier in recognizing subtle, nonstereotypical facial expressions. Emotion **17**, 1187–1198 (2017)
67. Tassinary, L.G., Cacioppo, J.T., Vanman, E.J.: The skeletomotor system: surface electromyography. Handbook of psychophysiology, vol. 3, 3rd edn., pp. 267–299 (2007)
68. Garrity, L.I.: Electromyography: a review of the current status of subvocal speech research. Mem. Cognit. **5**, 615–622 (1977)
69. Jacobson, E.: Electrophysiology of mental activities. Am. J. Psychol. **44**, 677–694 (1932)
70. Russell, J.A.: A circumplex model of affect. J. Pers. Soc. Psychol. **39**, 1161 (1980)
71. Barea, R., Boquete, L., Mazo, M., López, E.: System for assisted mobility using eye movements based on electrooculography. IEEE Trans. Neural Syst. Rehabil. Eng. **10**, 209–218 (2002)
72. Tamura, H., Murata, T., Yamashita, Y., Tanno, K., Fuse, Y.: Development of the electric wheelchair hands-free semi-automatic control system using the surface-electromyogram of facial muscles. Artif. Life Robot. **17**, 300–305 (2012)
73. Yan, M., Tamura, H., Tanno, K.: Gaze estimation using electrooculogram signals and its mathematical modeling. In: 2013 IEEE 43rd International Symposium on Multiple-Valued Logic, pp. 18–22 (2013).

74. Tamura, H., Yan, M., Sakurai, K., Tanno, K.: EOG-sEMG human interface for communication. Comput. Intell. Neurosci. **2016**, 7354082 (2016)
75. Williams, M.R., Kirsch, R.F.: Evaluation of head orientation and neck muscle EMG signals as command inputs to a human–computer interface for individuals with high tetraplegia (2008). https://doi.org/10.1109/tnsre.2008.2006216
76. Lu, Z., Chen, X., Zhang, X., Tong, K.-Y., Zhou, P.: Real-time control of an exoskeleton hand robot with myoelectric pattern recognition. Int. J. Neural Syst. **27**, 1750009 (2017)
77. Lu, Z., Chen, X., Zhao, Z., Wang, K.: A prototype of gesture-based interface. In: Proceedings of the 13th International Conference on Human Computer Interaction with Mobile Devices and Services, pp. 33–36. Association for Computing Machinery, New York, NY, USA (2011)
78. Khokhar, Z.O., Xiao, Z.G., Menon, C.: Surface EMG pattern recognition for real-time control of a wrist exoskeleton. Biomed. Eng. Online. **9**, 41 (2010)
79. Lu, Z., Zhou, P.: Hands-free human-computer interface based on facial myoelectric pattern recognition. Front. Neurol. **10**, 444 (2019)
80. Norman, D.: Emotion & design: attractive things work better. Interactions **9**, 36–42 (2002)
81. Mandryk, R.L., Inkpen, K.M., Calvert, T.W.: Using psychophysiological techniques to measure user experience with entertainment technologies. Behav. Inf. Technol. **25**, 141–158 (2006)
82. Mandryk, R.L., Atkins, M.S.: A fuzzy physiological approach for continuously modeling emotion during interaction with play technologies. Int. J. Hum. Comput. Stud. **65**, 329–347 (2007)
83. Guo, Y., Xia, Y., Wang, J., Yu, H., Chen, R.-C.: Real-time facial affective computing on mobile devices. Sensors **20** (2020). https://doi.org/10.3390/s20030870
84. Bao, Y., Cheng, Y., Liu, Y., Lu, F.: Adaptive Feature Fusion Network for Gaze Tracking in Mobile Tablets (2021). http://arxiv.org/abs/2103.11119

Chapter 3
Eye Movement

3.1 Movement of the Eyes

The eyeball is moved by the oculomotor muscles. There are four rectus muscles (the superior, inferior, lateral, and medial) and two oblique muscles (the superior and inferior). Working in pairs, the muscles are responsible for moving the eyeball in specific directions; six muscles are able to move the eyeball along each of the three axes. Seeing is a dynamic process; the human brain continually analyzes the visual field. The processes of visual attention enable the brain to decide where to look, and the eyes to follow that instruction by setting the necessary movement parameters.

Simultaneous movements of both eyes are differentiated according to their visual axes. The first type are vergence movements, in which the visual axes move in opposite directions; the second are conjugate movements, in which they move in the same direction. Vergences occur when an observer's gaze focuses on an object that is approaching or withdrawing from him/her while looking straight ahead. These can be either convergent or divergent. Such movements are controlled by the vergence movement system, which comprises the accommodative, proximal, tonic, and fusional subsystems [1]. The accommodative subsystem relates directly to the process of accommodation; a stimulus to accommodation is also one that leads the eyes to the convergent position. The proximal subsystem is connected with awareness of the distance between the observer and the object observed. The tonic subsystem pertains to muscle tone. The fusional subsystem integrates the information obtained from the others, and attempts to compensate for inaccuracies.

The following types of conjugate movements can be identified: tremor (mild eye twitching), drift, microsaccade, vestibulo-ocular reflex and optokinetic response, smooth pursuit movement, and saccadic movement (saccades). Fixations are treated as a separate category that consists of tremor, microsaccades, and drift. Although, physiologically, fixation is movement, it is interpreted as fixation of the gaze on a single element of an image—which, in itself, is a static process. Parameters of the

C. Biele, *Human Movements in Human-Computer Interaction (HCI)*,
Studies in Computational Intelligence 996,
https://doi.org/10.1007/978-3-030-90004-5_3

movements that form fixations, such as microsaccades, may, however, be analyzed and provide useful insights into mental processes.

Tremor is a wave-like motion of the eyes, with a frequency of approximately 90 Hz, which reflects the noise resulting from the base neural activation [2] and from imprecise control of the eye muscles [3]. Drift is a slow motion of the eye that moves the point of vision from the point of fixation in a centrifugal manner. The role of drift is to prevent decreases in the sensitivity of visual receptors (adaptation) under the action of a constant stimulus. It also compensates for the visual disturbances that can be caused by the shadows cast by blood vessels in the eyeball [4].

Microsaccades are small-magnitude, rapid, saccadic movements. They allow a viewer's gaze to move back to the point of fixation (a correction function resulting from small-magnitude eye movements) and maintain fixation, as a result. Due to minor instabilities of the visual system, a mechanism that allows the eye to refocus on the object being observed is indispensable for proper vision. Some studies show that microsaccades are important in the perception of color shades [2]. Other functions of the microsaccades include prevention of visual fading and the exploration of fine spatial detail [5].

Saccades are believed to be the fastest movements human bodies can make (they usually last no longer than 80 milliseconds). They typically occur between points of fixation. These movements are performed on average three times per second, which corresponds to the time necessary to process visual information at the point of fixation and to plan the next movement. The first step in the process of performing a saccade is switching attention. The decision to change the focus of attention is made by the structures of the parietal cortex [6]. At a later stage, the structures of the frontal lobe decide on the moment the eye should move and the nature of that movement. When a new stimulus appears in the visual field, the structures of the temporal-parietal-occipital association area in the right hemisphere are activated. Their information is integrated by the superior colliculi of the tectal plate, which has been christened the ultimate eye movement center [7]. These structures convey information to the brain stem, which coordinates the muscles responsible for the movement of the eyeballs.

Smooth pursuit movements allow moving objects to be tracked while the head remains almost motionless. Such movements block the saccades [8]. In natural conditions, tracking an object requires movement of the head and, occasionally, corrective saccades. Object tracking frequently involves saccades—meaning that their planning, execution, and control are subject to common mechanisms [9].

Stabilization movements, such as the vestibulo-ocular reflex (VOR) and the optokinetic response (OKR), can also be observed in the human eye. The VOR is a conjugate eyeball movement in the opposite direction of head movement. Its purpose is to stabilize the image on the retina when the head is moved rapidly—when it turns right or left, for instance. The VOR is triggered by acceleration, rather than constant motion. The reflex has a very short latency time of approximately 15 milliseconds, due to which it is possible to maintain a stable image on the retina even during rapid head movements, such as those that might occur while running. Importantly, during rotation of the head, the eyeballs move smoothly only until they reach their maximum angle of rotation; they will then move back to their primary position [10].

Another type of stabilization movement is the OKR, which also aims to stabilize images on the retina. As the information about the movement comes, in this case, from the visual system, the latency time of eye movement is much longer, approximately ranging between 50 and 100 milliseconds. As a result, the OKR does not compensate for rapid head movements, but complements the VOR during slow and steady ones. This occurs when excitation in the vestibular system is insufficient. The same happens during environmental movement in relation to a stationary observer—for example, when observing the landscape through the window of a moving train [11].

3.2 What is Eye Tracking?

The human visual field extends to 180 degrees horizontally and 130 degrees vertically. The details of images can only be seen, however, in a small area of the retina, the size of which is approximately two degrees. The eye muscles enable movements that enable projections of an image's key elements onto that area. The term, eye tracking is used to determine the direction of a user's gaze, which, in most cases, allows the identification of the object on which the eyes are fixed. With knowledge of the coordinates on the computer screen being observed, it is possible to determine what object was displayed in a given area at a given moment. Eye-tracking systems are available not only for standard stationary computer sets that display stimuli on monitor screens, but also for mobile solutions [12] and virtual reality (VR) headsets [13]. Such solutions, however, are much more complex—they must have a greater number of degrees of freedom, allow the head to move, and not assume that it remains motionless, as is the case of stationary eye-tracking devices. In addition to the points of gaze, eye trackers usually provide additional information about the eye, such as the size of the pupil or the time and frequency of blinking. These pieces form a valuable whole, both in research on human-computer interaction and, more broadly, in all studies that use computers to present content to subjects—such as those frequently conducted in psychology.

3.3 Use of Eye Movement Research in Human-Computer Interaction

As far as human-computer interaction is concerned, there are several areas in which eye movement research is used. A short description of each is presented below.

3.3.1 Usability Testing

The work of Fitts et al. [14] in the 1950s is considered the first usability study that tracked eye movements. The aim of their research was to discover the optimal locations of the instruments in airplane cockpits. Testing new devices, websites, or mobile applications that record user's eye movements enables identification of the areas on which the users' eyes fall when performing typical tasks—for instance, where users would expect to find a 'login' button [15].

3.3.2 Accessibility

Another area in which eye trackers are used is accessibility. People with spinal injuries or health conditions that cause loss of motor control can interact with their environments using eye-tracking systems to control computers and other devices.

3.3.3 Psychological Research

Tracking eye movements during computer-based experiments can provide highly valuable data in psychological research on attention, vision, cognitive functioning, and emotions. Eye tracking can also be used in psychology for diagnostic purposes [16].

3.3.4 Gaze-based Interaction

In recent years, work on eye-tracking as a method of controlling human-computer interaction has advanced significantly, and its prospects for further development are highly promising. In the second half of the 2010s, technology giants, such as Facebook, Google, and Apple acquired or invested in producers of eye-tracking software. The introduction of VR devices, such as the Vive Pro Eye, has also contributed to the popularization of eye tracking among consumers.

3.3.5 Market Research

Market research is an area in which eye tracking is used on a large scale. Its usability has been praised by designers of packaging and advertising materials [17]. By showing several versions of packaging to prospective clients, it is possible to compare how well and how quickly their key elements—such as the manufacturer's logo or name—are perceived, and how long they are able to maintain consumers' visual attention. The retail industry currently stocks tens of thousands of products in stores. From the perspective of producers (mainly those operating in the FMCG sector), it is crucial that their products are noticed by customers; this is only possible they catch the eye among those offered by the competition. Eye tracking enables the conduction of so-called "shelf studies" to determine which packaging variants stand out the most. Such studies involve presenting shelves on computer screens, in VR, or in the physical environment of a store [13, 18].

3.3.6 Authentication

Eye tracking is also used as an identification tool. Apart from the iris, other unique features of every human include the rapidity of eye movement and the dynamics of changes in the size of the pupil. Bednarik et al. [19] achieved identification efficiency of 60% when analyzing one-second fragments of eye movement data. Based on video recordings, Komorgortsev et al. [20] proposed the creation of the oculomotor plant model: a system comprising the eyeball, the muscles that move it, and brain signals. The authors determined the solution's quality using the half total error rate (HTER), which is defined as the mean of the false acceptance rate (FAR) and false rejection rate (FRR) [21]. Its maximum value amounted to 19%.

3.4 Eye Movements and Cognitive Processing

Until recently, eye movements were most often viewed physiologically, rather than psychophysiologically or psychologically. Recent enhancements in eye tracking, however, enable precise reflection of the interaction between cognitive processes and external visual stimuli. Some researchers even see eye tracking as a window to the mind, thoughts, and feelings [22]. Similar ideas—such as the eye-mind hypothesis—have been formulated in the past [23], in addition to numerous studies confirming the relationship between eye movement and cognitive function [24]. One construct used to measure eye-tracking activity is the Cognitive Load Theory (CLT) [25], which continues to play an integral role in research on human-computer interaction. Researchers recognize the need for non-invasive measurement of individual cognitive load; they believe that this can help designers to create interactive systems that do not "overload" users. Such systems gather data on cognitive load in real time—allowing them, for example, to increase or decrease the level of difficulty of a task (as is the case in e-learning systems) [26], or to adapt systems of critical importance to the cognitive states of users [27]. Examples of the use of the CLT cover a wide range of applications, including flight safety [28], human-centered design [29], human cognition modeling, usability, and multimedia learning. Reliable measurement of the cognitive load in real time, if possible, is important. Unfortunately, only a handful solutions that have adopted this approach are currently available [30].

The most commonly used indicators of cognitive load used in eye tracking include the duration of fixation, the speed of saccades, and changes in the size of the pupils. Indicators such as avoidance of looking at objects that demand more cognitive effort are used less frequently [3], and are applied solely in experimental comparisons. Unlike typical psychological tests, they lack diagnostic value due to the absence of standardization rules. As a consequence, it is impracticable to determine threshold values for such indicators. Various measures of cognitive load are briefly outlined below.

3.4.1 Fixations

The duration of fixation is related to the depth of visual information processing. Fixations that shorten over time can be considered an effect of cognitive fatigue—resulting in avoidance of deeper acquisition of new information (lower cognitive engagement). Just and Carpenter [31] demonstrated that during a variety of cognitive tasks, the duration of fixation depends on the type of cognitive operations. In their studies on reading, they also demonstrated that the eye stops when information needs to be processed i.e.—when reader draws conclusions at the end of a sentence. Such studies led to the formulation of the so-called eye-mind assumption—according to which whatever stimulus the eye fixates on, the mind processes.

3.4.2 Saccades

The speed of saccades is also considered an indicator of cognitive fatigue [32]. The longer the duration of tasks that require cognitive effort, the lesser the speed of saccades. The speed also decreases during tasks of higher difficulty [33, 34].

3.4.3 Changes in the Size of Pupils

Another eye tracking index associated with cognitive effort is pupil width. In the 1960s, Kahneman and Beatty [35] suggested that the diameter of the pupils was an effective indicator of subjects' momentary effort when performing cognitive tasks; Hess and Polt [36] demonstrated that changes in the size of the pupil related directly to the difficulty of mathematical tasks; later, Hyönä et al. [37] recorded changes in the size of the pupils during tasks of varying difficulty, and observed that that complicated ones, such as translating difficult words or repeating words in a foreign language, resulted in greater pupil dilation. Analogously, contraction of the pupils is most often associated with lower cognitive engagement in processing and acquiring new information. A more detailed analysis of the relationship between the size of the pupil and cognitive processing can be found in the review by van der Wel and van Steenbergen [38]. It is also noteworthy that the size of the pupil can also depend on a number of factors not related to cognitive load, such as the available light [39] and off-axis aberrations [40]. The literature proposes solutions to mitigate these types of problem [41, 42].

3.4.4 Gaze Aversion

Although used relatively rarely, the visual avoidance of a stimulus requiring increased cognitive effort can serve as a useful indicator. Both adults' and children's eyes are more likely to wander when answering difficult questions—which, interestingly, has been shown to improve memory [43, 44]. Looking away serves humans by reducing the amount of information arriving from outside and avoiding distractions [45]; this, in turn, facilitates the allocation of more cognitive resources to information processing [46].

3.4.5 Microsaccades

Microsaccades, tremor, and drift occur during fixation [47]. The relationship between microsaccade parameters and the difficulty of tasks was demonstrated by Siegenthaler et al.[48], who measured microsaccades during the performance of non-visual arithmetic tasks that were set at two levels of complexity. The study demonstrated that microsaccade rates decreased and microsaccade magnitudes increased when task difficulty increased. This effect is likely related to increased working memory load: in the case of high loads, the resources available for the performance and maintenance of fixations decrease, which explains their lower rates and reduced control over their execution (lower accuracy means greater amplitude). Gao et al. proved a similar relationship between the difficulty of tasks and microsaccade parameters [49]. A direct relationship between the frequency of saccades and the intensity of working memory load was demonstrated by Dalmaso et al. [50]. Analyses of the relationship between the difficulty of tasks and the magnitude of the saccades has shown that the difficulty of tasks might have accounted for 16% of saccade magnitude variance [51].

3.5 Cognitive Processing and Digital Reading

As technology advances, interactive and multimedia materials are enjoying increased popularity in education, and many countries consider them the primary form of content presentation in the teaching process. The advantages of such materials include visual attractiveness, the use of multiple communication channels (such as sight, hearing, and motor skills), and interaction. There are a host of reports [52] on the effectiveness of this approach that cover areas as diverse as arithmetic [53], history [54], introductory physics for engineers [55, 56], and electromagnetism [57]. There are also studies, however, that identify significant increases in the cognitive costs of learning with the use of such materials [58, 59]. Lower effectiveness of multimedia in education might be related to excesses of information, causing the cognitive system to become overloaded; this, in turn, might cause difficulties in the structuring or selection of the information obtained. Such problems can appear, for example, in the case of digital reading—particularly with materials that use hypertext [60, 61].

Hypertext is a means of organizing information that allows free movement between pieces of content, without the need for a predefined structure. Users can decide on the order in which hypertext materials will be read. The vast majority of current educational materials rely on this method of presentation. While enhancing interaction and improving the attractiveness of materials, it can also lead to increased cognitive demand [62].

The differences in cognitive load when learning from hypertext and traditional materials were studied by Krejtz et al. [63]. They used eye-tracking visual attention indicators related to the different levels of cognitive engagement that occurred among two groups of students: one that read hypertext, and another that read text presented in a traditional, linear manner. As expected, they discovered systematic differences in the eye-tracking indicators in the case of hypertext learning. One distinction could be observed in the presence of visual attention aversion from the material, which was more frequent in the hypertext group. Readers also exhibited decreases in the speed of saccades, the duration of fixation, and the size of the pupil—each of which varied while the students read subsequent stages of the text. Compared to linear text, hypertext demands greater engagement in structuring information and in making decisions regarding the selection of subsequent pieces of content [64]. This means that reading hypertext is associated with a greater external cognitive load. This might force greater cognitive effort, which, over time, contributes to decreases in cognitive engagement when acquiring new information. As a result, hypertext causes readers to be more tired than the traditional text. This explanation is consistent with the results of studies on the relationship between the size of the pupil and fatigue. Geacintov and Peavler [65] demonstrated that pupil constriction in the context of the work they performed might indicate fatigue. Morad et al. discovered that pupil size correlated negatively with subjectively assessed fatigue and sleepiness [66].

3.6 Eye Tracking for Human-Computer Interaction

Interfaces that analyze points of gaze seem innovative, and hold advantages over those that utilize mice or keyboards. Those with disabilities who can only move their eyes use such systems. Although they work well and prove highly useful to people who need them, their use has not become widespread among users whose physical condition does not necessitate it. Potential advantages of eye control systems include reduced strain of the arm and hand muscles compared with operating mice and keyboards, the potential for faster interaction (particularly when combined with other methods), and that cleaning and disinfection are unnecessary—which is of particular relevance during the ongoing COVID-19 pandemic. Another benefit is remote control. Zoom lenses and high-resolution cameras allow detection of the eye's position and point of gaze from a distance of several meters. Eye control, in its current state, nevertheless, remains imperfect. One of the most frequently cited disadvantages is the so-called Midas touch problem (see below). Others lie in humans' inability to consciously control certain eye movements, and in eye muscle fatigue.

3.6.1 Using Eye Movements to Control Interaction

Human-computer interaction that uses sight has been the subject of empirical research for over 40 years. It has been studied in many applications, including simulators [67], arcade games [68], platform adventure games [69], and massive multiplayer online games [70]. Jacob [71], and Starker and Bolt [72] are considered the first to have attempted to use eye movements in the modeling of human-computer interactions. Both studies describe the use of the dwell time measure—either to trigger desired actions in controlled systems, or as a measure of interest in elements on which users would fix their gaze. One of the major difficulties in eye-controlled interaction is the Midas touch problem. Jacob was the first to draw attention to it, and to the potential of dwell time in eliminating it. In most interactions with computers, control is indirect—for example, the actions of characters in video games are controlled by moving and clicking a mouse. The scanning of a screen by a player's eyes is a similarly important element of human-computer interaction. Scanning of the screen can be misinterpreted as significant eye movement, and lead to unwanted user actions. With that in mind, scientists agree that eye-controlled interaction cannot simply replace pointing devices, such as the mouse, and that implementing eye control in an application usually requires it to be redesigned. This is due to fundamental differences between the mechanics of controlling interfaces with the eyes, with touch, or with a mouse. In the case of the mouse, which is the most widespread pointing device used in interactions with computers, there are at least two modes of operation. The first is used for pointing, and the second for selecting. These modes are switched using buttons. Achieving comparable functionality for eye control is rather complicated—although it has not discouraged researchers interested in the subject to propose different ways of switching modes with the eyes. The solutions include blinking and winking [73, 74], conscious pupil dilation and constriction [75], and dwell time [76]. One problem that is specific to eye movement interaction is the precision of cursor control. When a mouse is not moved, its cursor remains in the same position; conversely, physiological mechanisms prevent the eye from remaining entirely motionless. Additionally, the measurements of the position of the eye made by eye trackers is not very accurate. This leads to aiming errors and poor gaze precision during interactions with computers.

3.6.2 A potential Solution to the Midas Touch Problem

Krejtz et al. [77] tested gaze control in a simple maze game. They observed that players performed more poorly and enjoyed the game less during gaze-controlled sessions. This was likely due to the Midas touch problem—visual scanning with no intention of movement was interpreted as signals to initiate the character's movement in the game. The solution proposed centered on modifying the mechanics of the character's locomotion by means of a cut-off function: gaze control would be active only when the gaze was within at a certain distance of the character's position; eye fixations beyond this radius would be interpreted as visual scanning, and would not affect the character's movement. An alternative to the cut-off function involves controlling the speed of the game character using the Gaussian function. Its input would be the distance from the current gaze position to the game character. This approach was applied by Biele et al. [78], who tested an innovative gaze control method relying on a mechanism that prevented unintended user actions. Their experiment confirmed the effectiveness of gaze control using the Gaussian velocity transformation: the performance of the players was the highest of all the gaze-controlled solutions that were tested. The subjective ratings of the game were consistent with the level of players' performance: the gaze-controlled game with Gaussian transformation was rated the highest in terms of interaction and playability. The players found the game based on the Gaussian transformation enjoyable, natural, easy, and engaging. The results were comparable to the ratings of the keyboard-controlled version of the game. Analyses of the distribution of the points of gaze clearly demonstrated that the Gaussian function facilitated visual scanning. The number of gazes at the peripheral parts of the screen (scanning) was the highest in the Gaussian transformation game.

3.6.3 Error Correction in Head-Mounted Eye Trackers

Despite eye-tracking technology having existed for many years, the error resulting from the accuracy of registration offered by the devices available remains a serious concern. This is particularly noticeable in mobile devices. The first issue to be settled when working with eyeball movement is correct mapping of the point of gaze based on video signals containing images of the eye. The accuracy of gaze point estimation is one of the main factors limiting the development of applications that could utilize gaze control. This is particularly evident in the case of head-mounted eye trackers, which allow users to observe screens from off-center positions. Other errors are generated by screen-tracking algorithms that use IR LEDs or QR codes, and head-mounted cameras that observe the environment to locate the screen within the user's visual field. In the systems used at present, the quality of algorithms and procedures for internal eye tracking calibration is very high. There are no solutions that allow estimation of errors caused by the specificity of screen position detection algorithms in the user's visual field, however.

To address this problem, a solution [79] was proposed to eliminate errors in gaze-based multiuser interactions. It relies on the surface recalibration—a procedure used to estimate the errors resulting from the inaccuracy of the algorithms that track the position of a screen, caused, for example, by the user moving away from its center. The data received from the eye tracker is transformed in real time with the use of data collected during the surface recalibration procedure. The procedure primarily involves collecting data on gazes at a series of points on the screen. Each point is displayed for three seconds, during which data on gaze position is recorded. Based on this, the recalibration errors necessary for real-time data processing are calculated. Each gaze data point then undergoes a two-stage online transformation: first, for each of the surface dimensions (X, Y), a weight vector of length n is calculated (where n equals the number of recalibration points); then, the gaze position is corrected by the sum of n weighted recalibration errors. This operation is equivalent to calculating the difference between the gaze position coordinate and the dot product of two vectors: that of successive recalibration errors and of the weights obtained during the previous step. In order to test the method, its authors conducted an empirical study that compared the effectiveness of gaze interaction systems with and without the recalibration function. The efficiency of the system with the recalibration function was expected to be higher; meaning that the task performed with the use of the recalibration system would be executed faster. Subjects were asked to point at a series of circular targets arranged concentrically on a screen with a gazed-controlled cursor. To complete the task, they were required to mark the target by hovering the cursor on it. After marking all targets, another circle appeared. The whole process consisted of five circles.

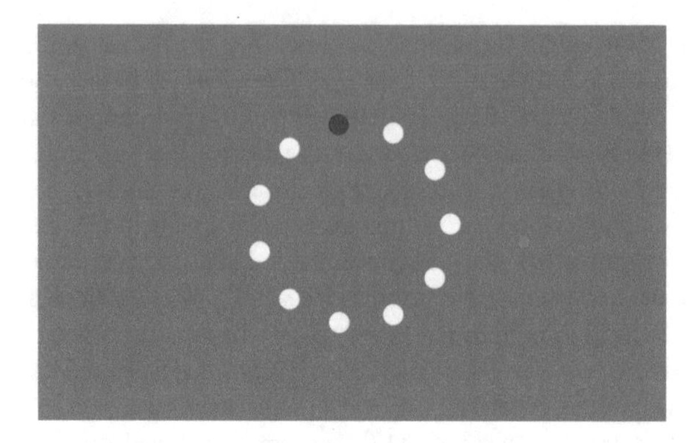

The subjects' reaction times (the time needed to mark the target) were then analyzed. The analyses demonstrated a statistically significant difference: the reaction time was approximately 1.5 sec without and 1.25 sec with recalibration.

References

1. Von Noorden, G.K., Campos, E.C.: Binocular Vision and Ocular Motility: Theory and management of strabismus 6th edn. A Harcourt Health Sciences Company (2002)
2. Martinez-Conde, S., Macknik, S.L., Hubel, D.H.: The role of fixational eye movements in visual perception. Nat. Rev. Neurosci. **5**, 229–240 (2004)
3. Holmqvist, K., Nyström, M., Andersson, R., Dewhurst, R., Jarodzka, H., Van de Weijer, J.: Eye Tracking: A Comprehensive Guide to Methods and Measures. OUP Oxford (2011)
4. Sekuler, R., Blake, R.: Perception. McGraw-Hill (1994)
5. Poletti, M., Rucci, M.: A compact field guide to the study of microsaccades: challenges and functions. Vision Res. **118**, 83–97 (2016)
6. Thiel, C.M., Zilles, K., Fink, G.R.: Cerebral correlates of alerting, orienting and reorienting of visuospatial attention: an event-related fMRI study. Neuroimage **21**, 318–328 (2004)
7. Liversedge, S.P., Findlay, J.M.: Saccadic eye movements and cognition. Trends Cogn. Sci. **4**, 6–14 (2000)
8. Ross, J., Morrone, M.C., Burr, D.C.: Compression of visual space before saccades (1997). https://doi.org/10.1038/386598a0
9. Srihasam, K., Bullock, D., Grossberg, S.: Coordinating saccades and smooth pursuit eye movements during visual tracking and perception of objects moving with variable speeds. J. Vis. **6**, 908 (2006)
10. Schweigart, G., Mergner, T., Evdokimidis, I., Morand, S., Becker, W.: Gaze stabilization by optokinetic reflex (OKR) and vestibulo-ocular reflex (VOR) during active head rotation in man. Vision Res. **37**, 1643–1652 (1997)
11. Krauzlis, R.J.: Eye movements (2013). https://doi.org/10.1016/b978-0-12-385870-2.00032-9
12. Stuart, S., Hunt, D., Nell, J., Godfrey, A., Hausdorff, J.M., Rochester, L., Alcock, L.: Do you see what I see? mobile eye-tracker contextual analysis and inter-rater reliability. Med. Biol. Eng. Comput. **56**, 289–296 (2018)
13. Meißner, M., Pfeiffer, J., Pfeiffer, T., Oppewal, H.: Combining virtual reality and mobile eye tracking to provide a naturalistic experimental environment for shopper research. J. Bus. Res. **100**, 445–458 (2019)
14. Fitts, P.M., Jones, R.E., Milton, J.L.: Eye movements of aircraft pilots during instrument-landing approaches. Ergon. Psychol Mech. Models Ergon. **3**, 56 (2005)

15. Pan, B., Hembrooke, H.A., Gay, G.K., Granka, L.A., Feusner, M.K., Newman, J.K.: The determinants of web page viewing behavior: an eye-tracking study. In: Proceedings of the 2004 Symposium on Eye Tracking Research & Applications, pp. 147–154. Association for Computing Machinery, New York, NY, USA (2004)
16. Brunyé, T.T., Drew, T., Weaver, D.L., Elmore, J.G.: A review of eye tracking for understanding and improving diagnostic interpretation. Cognit. Res. Princ. Implic. **4**, 7 (2019)
17. Moskowitz, H.R., Moskowitz, H., Lawlor, J.B., Gupton, A., Deliza, R.: Food Packaging Research and Consumer Response. Blackwell Publishing Professional (2007)
18. Tonkin, C., Ouzts, A.D., Duchowski, A.T.: Eye tracking within the packaging design workflow (2011). https://doi.org/10.1145/1983302.1983305
19. Bednarik, R., Kinnunen, T., Mihaila, A., Fränti, P.: Eye-Movements as a Biometric. In: Image Analysis, pp. 780–789. Springer Berlin Heidelberg (2005)
20. Komogortsev, O.V., Karpov, A., Price, L.R., Aragon, C.: Biometric authentication via oculomotor plant characteristics. In: 2012 5th IAPR International Conference on Biometrics (ICB), pp. 413–420 (2012)
21. Jain, A.K., Pankanti, S., Prabhakar, S., Hong, L., Ross, A.: Biometrics: a grand challenge. In: Proceedings of the 17th International Conference on Pattern Recognition, 2004. ICPR 2004, vol. 2, pp. 935–942 (2004)
22. Glimcher, P.W.: The neurobiology of visual-saccadic decision making. Annu. Rev. Neurosci. **26**, 133–179 (2003)
23. Just, M.A., Carpenter, P.A.: A theory of reading: from eye fixations to comprehension. Psychol. Rev. **87**, 329–354 (1980)
24. Zee, D.S., Lasker, A.G.: Antisaccades: probing cognitive flexibility with eye movements (2004)
25. Sweller, J.: Cognitive load during problem solving: effects on learning. Cogn. Sci. **12**, 257–285 (1988)
26. Rakoczi, G., Duchowski, A., Pohl, M.: designing online tests for a virtual learning environment: evaluation of visual behaviour at different task types. In: Proceedings of the International Conference on Human Behavior in Design (HBiD). The Design Society, Ascona, Switzerland (2014)
27. Shi, Y., Choi, E., Taib, R., Chen, F.: designing cognition-adaptive human-computer interface for mission-critical systems. In: Papadopoulos, G.A., Wojtkowski, W., Wojtkowski, G., Wrycza, S., Zupancic, J. (eds.) Information Systems Development: Towards a Service Provision Society, pp. 111–119. Springer, US, Boston, MA (2010)
28. Peysakhovich, V.: Study of pupil diameter and eye movements to enhance flight safety (2016)
29. Oviatt, S.: Human-centered design meets cognitive load theory: designing interfaces that help people think. In: Proceedings of the 14th ACM International Conference on Multimedia, pp. 871–880. Association for Computing Machinery, New York, NY, USA (2006)
30. Yuksel, B.F., Oleson, K.B., Harrison, L., Peck, E.M., Afergan, D., Chang, R., Jacob, R.J.K.: Learn Piano with BACh: An adaptive learning interface that adjusts task difficulty based on brain state. In: Proceedings of the 2016 CHI Conference on Human Factors in Computing Systems, pp. 5372–5384. Association for Computing Machinery, New York, NY, USA (2016)
31. Just, M.A., Carpenter, P.A.: Eye fixations and cognitive processes. Cogn. Psychol. **8**, 441–480 (1976)
32. Di Stasi, L.L., Catena, A., Cañas, J.J., Macknik, S.L., Martinez-Conde, S.: Saccadic velocity as an arousal index in naturalistic tasks. Neurosci. Biobehav. Rev. **37**, 968–975 (2013)
33. Di Stasi, L.L., Antolí, A., Cañas, J.J.: Evaluating mental workload while interacting with computer-generated artificial environments. Entertain. Comput. **4**, 63–69 (2013)
34. Di Stasi, L.L., Renner, R., Staehr, P., Helmert, J.R., Velichkovsky, B.M., Cañas, J.J., Catena, A., Pannasch, S.: Saccadic peak velocity sensitivity to variations in mental workload. Aviat. Space Environ. Med. **81**, 413–417 (2010)
35. Kahneman, D., Beatty, J.: Pupil diameter and load on memory. Science **154**, 1583–1585 (1966)
36. Hess, E.H., Polt, J.M.: Pupil size in relation to mental activity during simple problem-solving. Science **143**, 1190–1192 (1964)

37. Hyönä, J., Tommola, J., Alaja, A.M.: Pupil dilation as a measure of processing load in simultaneous interpretation and other language tasks. Q. J. Exp. Psychol. A. **48**, 598–612 (1995)
38. van der Wel, P., van Steenbergen, H.: Pupil dilation as an index of effort in cognitive control tasks: a review. Psychon. Bull. Rev. **25**, 2005–2015 (2018)
39. Beatty, J., Lucero-Wagoner, B., Others: The pupillary system. Handbook of Psychophysiology, vol. 2 (2000)
40. Mathur, A., Gehrmann, J., Atchison, D.A.: Pupil shape as viewed along the horizontal visual field. J. Vis. **13** (2013). https://doi.org/10.1167/13.6.3
41. Hayes, T.R., Petrov, A.A.: Mapping and correcting the influence of gaze position on pupil size measurements. Behav. Res. Methods. **48**, 510–527 (2016)
42. Raiturkar, P., Kleinsmith, A., Keil, A., Banerjee, A., Jain, E.: Decoupling light reflex from pupillary dilation to measure emotional arousal in videos. In: Proceedings of the ACM Symposium on Applied Perception, pp. 89–96. Association for Computing Machinery, New York, NY, USA (2016)
43. Doherty-Sneddon, G., Bruce, V., Bonner, L., Longbotham, S., Doyle, C.: Development of gaze aversion as disengagement from visual information. Dev. Psychol. **38**, 438–445 (2002)
44. Glenberg, A.M., Schroeder, J.L., Robertson, D.A.: Averting the gaze disengages the environment and facilitates remembering. Mem. Cognit. **26**, 651–658 (1998)
45. Ehrlichman, H.: From gaze aversion to eye-movement suppression: An investigation of the cognitive interference explanation of gaze patterns during conversation. Br. J. Soc. Psychol. **20**, 233–241 (1981)
46. Doherty-Sneddon, G., Phelps, F.G.: Gaze aversion: a response to cognitive or social difficulty? Mem. Cognit. **33**, 727–733 (2005)
47. Engbert, R., Kliegl, R.: Microsaccades uncover the orientation of covert attention. Vision Res. **43**, 1035–1045 (2003)
48. Siegenthaler, E., Costela, F.M., McCamy, M.B., Di Stasi, L.L., Otero-Millan, J., Sonderegger, A., Groner, R., Macknik, S., Martinez-Conde, S.: Task difficulty in mental arithmetic affects microsaccadic rates and magnitudes. Eur. J. Neurosci. **39**, 287–294 (2014)
49. Gao, X., Yan, H., Sun, H.-J.: Modulation of microsaccade rate by task difficulty revealed through between- and within-trial comparisons. J. Vis. **15** (2015). https://doi.org/10.1167/15.3.3
50. Dalmaso, M., Castelli, L., Scatturin, P., Galfano, G.: Working memory load modulates microsaccadic rate. J. Vis. **17**, 6 (2017)
51. Krejtz, K., Duchowski, A.T., Niedzielska, A., Biele, C., Krejtz, I.: Eye tracking cognitive load using pupil diameter and microsaccades with fixed gaze. PLoS One. **13**, e0203629 (2018)
52. Paas, F., Van Gerven, P.W.M., Wouters, P.: Instructional efficiency of animation: effects of interactivity through mental reconstruction of static key frames. Appl. Cogn. Psychol. **21**, 783–793 (2007)
53. Vallée-Tourangeau, F.: Interactivity, efficiency, and individual differences in mental arithmetic (2013). https://doi.org/10.1027/1618-3169/a000200
54. Williams, T.: Multimedia learning gets medieval. Pedagogy **9**, 77–95 (2009)
55. Goodman, D., Rueckert, F., O'Brien, J.: initial steps toward a study on the effectiveness of multimedia learning modules in introductory physics courses for engineers. https://doi.org/10.18260/1-2-28531
56. Stelzer, T., Gladding, G., Mestre, J.P., Brookes, D.T.: Comparing the efficacy of multimedia modules with traditional textbooks for learning introductory physics content. Am. J. Phys. **77**, 184–190 (2009)
57. Moore, J.: Efficacy of multimedia learning modules as preparation for lecture-based tutorials in electromagnetism (2018). https://doi.org/10.3390/educsci8010023
58. Hegarty, M., Quilici, J., Narayanan, N.H., Holmquist, S., Moreno, R.: Multimedia instruction: lessons from evaluation of a theory-based design. J. Educ Multimed Hypermed. **8**, 119–150 (1999)
59. Boucheix, J.-M., Lowe, R.K.: An eye tracking comparison of external pointing cues and internal continuous cues in learning with complex animations. Learn. Instr. **20**, 123–135 (2010)

60. O'Hara, K., Sellen, A.: A comparison of reading paper and on-line documents. In: Proceedings of the ACM SIGCHI Conference on Human factors in computing systems, pp. 335–342. Association for Computing Machinery, New York, NY, USA (1997)
61. Fortunati, L., Vincent, J.: Sociological insights on the comparison of writing/reading on paper with writing/reading digitally (2014). https://doi.org/10.1016/j.tele.2013.02.005
62. Moreno, R., Mayer, R.: Interactive multimodal learning environments. Educ. Psychol. Rev. **19**, 309–326 (2007)
63. Krejtz, K., Biele, C., Jonak, L.: Visual attention dynamics and cognitive engagement during hypertextreading. Studia Psychologiczne (2015)
64. Khan, K., Locatis, C.: Searching through cyberspace: The effects of link display and link density on information retrieval from hypertext on the world wide web. J. Am. Soc. Inf. Sci. **49**, 176–182 (1998)
65. Geacintov, T., Peavler, W.S.: Pupillography in industrial fatigue assessment. J. Appl. Psychol. **59**, 213–216 (1974)
66. Morad, Y., Lemberg, H., Yofe, N., Dagan, Y.: Pupillography as an objective indicator of fatigue. Curr. Eye Res. **21**, 535–542 (2000)
67. Nielsen, A.M., Petersen, A.L., Hansen, J.P.: Gaming with gaze and losing with a smile. In: Proceedings of the Symposium on Eye Tracking Research and Applications, pp. 365–368. Association for Computing Machinery, New York, NY, USA (2012)
68. Dorr, M., Pomarjanschi, L., Barth, E.: Gaze beats mouse: a case study on a gaze-controlled breakout. PsychNology J. **7** (2009)
69. Muñoz, J., Yannakakis, G.N., Mulvey, F., Hansen, D.W., Gutierrez, G., Sanchis, A.: Towards gaze-controlled platform games. In: 2011 IEEE Conference on Computational Intelligence and Games (CIG'11), pp. 47–54 (2011)
70. Istance, H., Hyrskykari, A., Vickers, S., Chaves, T.: For your eyes only: Controlling 3D online games by eye-gaze (2009) https://doi.org/10.1007/978-3-642-03655-2_36
71. Jacob, R.J.K.: What you look at is what you get: eye movement-based interaction techniques. In: Proceedings of the SIGCHI Conference on Human Factors in Computing Systems, pp. 11–18. Association for Computing Machinery, New York, NY, USA (1990)
72. Starker, I., Bolt, R.A.: A gaze-responsive self-disclosing display. In: Proceedings of the SIGCHI Conference on Human Factors in Computing Systems, pp. 3–10. Association for Computing Machinery, New York, NY, USA (1990)
73. Ohno, T., Mukawa, N., Kawato, S.: Just blink your eyes: a head-free gaze tracking system. In: CHI '03 Extended Abstracts on Human Factors in Computing Systems, pp. 950–957. Association for Computing Machinery, New York, NY, USA (2003)
74. Špakov, O.: EyeChess: the tutoring game with visual attentive interface. Altern Access Feel. Game. **5** (2005)
75. Ekman, I.M., Poikola, A.W., Mäkäräinen, M.K.: Invisible eni. In: Proceeding of the 26th Annual CHI Conference Extended Abstracts on Human Factors in Computing Systems—CHI '08. ACM Press, New York, New York, USA (2008). https://doi.org/10.1145/1358628.1358820.
76. Bednarik, R., Gowases, T., Tukiainen, M.: Gaze interaction enhances problem solving: effects of dwell-time based, gaze-augmented, and mouse interaction on problem-solving strategies and user experience. J. Eye Mov. Res. **3** (2009). https://doi.org/10.16910/jemr.3.1.3
77. Krejtz, K., Biele, C., Chrzastowski, D., Kopacz, A.: Gaze-controlled gaming: immersive and difficult but not cognitively overloading. Proceedings of the 2014 ACM International Joint Conference on Pervasive and Ubiquitous Computing: Adjunct Publication (2014)
78. Biele, C., Chrząstowski-Wachtel, D., Młodożeniec, M., Niedzielska, A., Kowalski, J., Kobyliński, P., Krejtz, K., Duchowski, A.T.: Gaussian function improves gaze-controlled gaming (2018). https://doi.org/10.1007/978-3-319-59424-8_23
79. Biele, C., Kobylinski, P.: surface recalibration as a new method improving gaze-based human-computer interaction. In: Intelligent Human Systems Integration, pp. 197–202. Springer International Publishing (2018)

Chapter 4
Hand Movements Using Keyboard and Mouse

4.1 Introduction

Traditional human–computer interaction devices, which have been in use for several decades, are based on movement of human muscles. A user who is unable to move his/her fingers would find it impossible to type on a standard computer keyboard. Also to operate a mouse, the muscles of the hands and forearms must be used. Since the use of keyboards and mice is a manifestation of motor behavior, it is possible to apply these peripheral devices not only to the entering of text, but also to obtain information on psychological state of the users i.e. the their emotional states. As human movement patterns are unique, keyboards and mice can also be used to identify specific users. It is worth noting that the movements we make when typing on a keyboard compared to handwriting differ significantly. This affects how information handwritten or entered using a keyboard is processed, and is of particular relevance in the fields of remote learning and e-learning.

4.2 Keyboard

The contemporary keyboard dates back to the nineteenth century. Its predecessor, the typewriter was invented by Christopher Latham Sholes, who, alongside James Densmore, patented the first iteration of the device in 1868. Soon after, typewriters were commercialized on a large scale by Remington. In the ensuing decades, typewriters gradually evolved into keyboards, which have since then been used as a tool to assist in human–computer interaction. Today, the most common keyboard layout is "QWERTY" or variants thereof—a layout that can appear unnatural, or even puzzling to inexperienced keyboard users. A multitude of solutions have been proposed to address this issue, including the Dvorak Simplified Keyboard—a layout designed to be more efficient from the user's perspective. However, these alternative

C. Biele, *Human Movements in Human-Computer Interaction (HCI)*,
Studies in Computational Intelligence 996,
https://doi.org/10.1007/978-3-030-90004-5_4

layouts did not gain popularity. How is it possible that much more effective keyboard layouts (compared to QWERTY) have become of marginal importance? The most common explanation [1] for the standard keyboard layout is that it was designed to address the mechanical limitations of the typewriter. Normally, pressing a key would raise the corresponding type bar, which would then strike an ink ribbon to produce a character on a sheet of paper. Pressing several keys in rapid succession, however, often resulted in jamming of type bars due to their curvature. In recent years, this explanation has been questioned, and the available data indicates that the "QWERTY" keyboard layout was designed to facilitate the transfer of information in Morse code [2].

4.2.1 Keyboard Usage Dynamics and Identification

The present day, an era in which systems routinely store highly sensitive data, such as that related to banking, issues related to the verification and identification of users are becoming increasingly relevant. Unauthorized access to an online account can lead to loss of data, virtual resources, and, in the case of banking systems, also financial losses. Unauthorized access has the potential to cause significant damage even in relatively less critical systems, such as internet forums or computer systems in schools and universities. To prevent such incidents, specialists constantly develop new systems that do not rely on textual passwords, which could be stolen or cracked, or which are simply too weak [3]. Password-based systems also run the risk of unauthorized access due to inadvertent user actions, such as leaving their computers before logging out. No such risks exist, however, in biometric systems based on monitoring behavior patterns inferred from standard methods of interaction with computers, such as use of keyboards and mice. Such systems allow users to be verified on the basis of their identities rather than the knowledge they possess (passwords), the items they possess (identity cards), or combinations of the above (two-stage identification).

Keyboard dynamics is the name given to the unique time patterns related to the use of keyboards by specific users. The most frequently used parameters include the length of time keys are depressed and the length of time between presses. A solid body of research has been conducted pertaining to the potential use of keyboard dynamics in security systems as a tool for user identification. The dynamics of using a keyboard is adequately unique (as much so as a handwritten signature) to identify an individual. Work on the subject initially focused on strengthening standard passwords with a component based on keystroke dynamics [4]. The potential of the technology is relatively well researched [4–6]. Users are commonly identified on the basis of the dynamics inferred when entering their usernames and passwords. The methods used predominantly employ fixed-text models: they use the same piece of text both for training the model and for identifying users during the authentication process. There are also solutions that are able to use any text that is assigned to a specific user [4, 7, 8]. These methods, however, are less effective. Free-text models perform better when the text involved is longer, but this is impractical for identification purposes. Such methods require users to enter any text. Users' activity can also be monitored in the background, as long as their computers remain in use [7, 8]. Interestingly, these models can prove useful for those who lack fluency in the English language, as text can be entered in different languages—some in users' mother tongues, and some in English [9]. The advantage of this approach is that it is less intrusive to users, as data is gathered discreetly. Importantly, this reduces users' cognitive demand and mitigates the burden of interruptions that negatively affect cognitive functioning [10]. Various types of algorithms are used to identify users in keystroke dynamic analysis methods, such as neural networks [11], decision trees [12], and distance measurements [6].

4.2.2 Keyboard Usage Dynamics and Emotions

Developers of computer interfaces have long sought methods of collecting information on the emotional states of users. Such contextual information would allow the creation of a system that could adapt effectively to users' emotions. In recent years, multiple methods to this end have been developed, including those based on voice intonation analysis, facial expression analysis, skin sensors, and facial thermal imaging. Despite their effectiveness, these methods continue to demonstrate two major weaknesses: they are highly intrusive and burdensome for users; and they frequently require expensive, specialized equipment that is inaccessible in typical office environments. These difficulties have been addressed with the use of standard computer peripherals to measure emotions on the basis of behavior [13].

A particularly innovative approach to this was developed by Epp et al. [14]. They devised a solution that relied on detecting users' emotional states by analyzing the rhythms of their typing on an ordinary computer keyboard (keystroke dynamics [6, 15]). It appears promising in the detection of emotions in human–computer interaction. The analysis of keystroke dynamics and its use to identify users' emotional states are not negatively impacted by any of the aforementioned shortcomings. The entirely non-invasive method relies on standard computer hardware, and presents high potential for systems that possess access to data on user emotions. The authors developed 15 classifiers for various emotional states. They report that the most accurate classifier results (above 77%) were obtained for nervousness, sadness, fatigue, self-confidence, anger, and excitement. The data on emotions used for classification being sourced from self-reporting, however, presents a significant disadvantage.

4.2.3 The Keyboard for Note Taking

In recent years, the use of laptops (and by extension, keyboards) to take notes [16] has gained in popularity among university students during lectures and classes. While students have embraced the technology [17, 18], their teachers remain reluctant [19].

There are a host of reports available on the effectiveness of laptops on traditional note-taking in the literature [20, 21]. Taking notes during classes has a two-fold positive effect on the processing and memorizing of academic material—it improves both encoding and external storage functions. Since both functions are not mutually exclusive, they can occur in tandem. The literature distinguishes between two types of note-taking—generative and non-generative. Non-generative note-taking—in other words, copying the information presented—results only in shallow processing [22]. The deeper the information delivered is processed, the easier it is for students to remember. These statements are scientifically proven. Studies indicate that verbatim note-taking ("copying and pasting") leads to limitations in the absorption of knowledge—particularly in the case of more complex material [23]. Does this mean that students' use of computers and propensity to type in the classroom promote the verbatim note-taking strategy, which may, in turn, lead to less efficient absorption of the material presented by their teachers?

Mueller and Oppenheimer sought the answer to this question by conducting series of studies in which they compared laptop and longhand note-taking [24]. They were interested in how the manner of note-taking affected the memorization of material. Subjects were shown fifteen-minute lectures, during which they were requested to take notes. They were then quizzed on the subjects presented in the videos. The researchers discovered, both in the case of factual questions and of conceptual-application questions, that the participants who used laptops for note-taking performed significantly worse than those who had relied on handwritten notes. It is worth mentioning that although the differences were noticeable in the case of information that required processing, they were much less pronounced in the case of simpler information. These differences might be explained by the use of keyboards eliciting different note-taking strategies. Analyses of the notes taken showed that the students who used computers noted verbatim information provided during the lectures more frequently, despite having been instructed to avoid doing so. The keyboard may cause difficulties in noting information synthetically, in addition

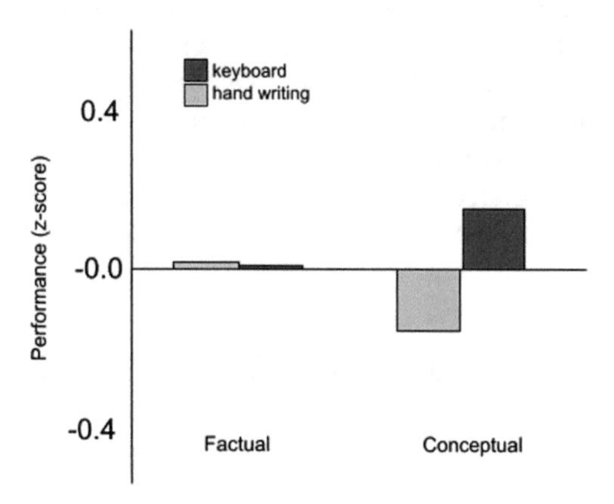

Fig. 4.1 Factual and conceptual data recall, adapted from [24]

to hindering the easy creation of graphic materials that efficiently summarize information. It also naturally disposes users to transcribe lectures, leading to shallower processing, which is detrimental to memorization [25]. Since handwriting is more efficient, it seems advisable to promote tablets and styluses as technology tools for note-taking. Unfortunately, detailed studies comparing how knowledge is absorbed during lectures with the use of electronic and traditional handwriting methods are scarce. The initial conclusions on this matter suggest, however, that even the use of tablets and styluses might prove insufficient. A recent study reports that individuals using traditional paper notepads for note-taking, compared to those using iPads and styluses, demonstrated increased activation in the areas of the brain responsible for memory, visual imagery, language processing, and information encoding [26]. The results demonstrate clearly that the use of traditional, paper notepads influences these higher-level mental functions. The increasing popularity of computers and tablets as assistive tools in learning is facilitating a decline in the use of paper notepads. Understanding of the impact of this change on the learning process, however, remains incomplete.

Note-taking, either with a traditional pen or with a tablet stylus, activates hand and arm muscles to a much greater extent than typing. Writing by hand is also a more intense sensory experience. Both of these observations suggest that handwriting and typing differ significantly in terms of brain stimulation. The results of the studies described above demonstrate that the simultaneous stimulation of channels when writing by hand (touch, motor skills, sight) facilitates the memorization of noted information. To write a character by hand, a complex motor pattern must be evoked. During the process of writing, continuous feedback is received from the visual channel—we have to control what we are writing. Neuroimaging studies have revealed that the brain has a network of interconnected areas responsible for handwriting [27], and that it is different to that activated during typing [28]. At the brain level, the act of reading handwritten text also differs [29] from reading printed text. The area of the motor cortex responsible for moving the hand is more active when reading handwritten text than when reading printed text. It probably stems from the fact that handwritten letters, despite their stacicity at the moment of perception, are the result of an action—they are the product of a moving hand. It is believed that this phenomenon is due to a mechanism similar to the implied motion perception [30] (see Chap. 9). Differences in activation have also been observed in the supplemental motor area, which is responsible, among other things, for the synchronization of movements in sequences, and whose role might be more integral than that of the visual cortex. Increased activations also occur in the frontal eye field, which is responsible, among other things, for saccades (swift eye movements to scan the visual field); and in the fusiform gyrus, which facilitates recognition of similar stimuli at an expert level—the most prominent example being human faces.

4.3 Mouse

The history of the computer mouse is much shorter than that of the keyboard. It dates back to the 1960s, when Douglas Engelbart developed a wooden pointing device comprising two metal wheels and a button [31]. Although the rudimentary mouse could move in all directions, that movement was far from smooth. When it was moved upward, for instance, the horizontal wheel applied friction to the surface. The name was coined by Engelbart himself, who admitted in several interviews that he often likened his wired device to a small rodent with a tail. Subsequent commercial versions of the device were fitted with up to three buttons, and the wheels were superseded by rollers and balls. The moving parts were swapped for optical sensors in the early 1980s. Eventually, the mouse lost its tail, and most mice are presently wireless. Despite the plethora of mouse variants on the market—some of which offer innovative solutions, such as gesture recognition (Apple mice), or snapping flat (Microsoft Arc)—the modern computer mouse continues to resemble its prototype. Although alternative pointing devices, like touchpads or trackballs, attempt to compete with the mouse, it remains among the most popular computer peripherals.

4.3.1 Mouse Movement and Identification

As with the keyboard, behavior parameters pertaining to the use of mice can be used to identify users. Currently, the most popular verification and authentication methods require specific actions to be taken by users. These methods verify or authenticate users only once, which leaves them vulnerable to threats, such as external actors taking control of users' sessions. A system proposed by Zheng et al. [32], which relies on the analysis of mouse movements, is immune to such threats, however. The requirements for an identification system based on the analysis of mouse movements that runs constantly in the background are rather high. Systems must operate quickly and their number of false positives—situations in which users are incorrectly identified as intruders—must be reduced virtually to zero. Compared to keystroke dynamic identification, mouse movement monitoring systems possess an inherent advantage in that they require no access to keyboard data (which, in turn, might contain sensitive data). In addition, when performing certain activities on computers (such as web surfing), keyboards are used sparingly—meaning that keystroke dynamic identification might fail to work as intended.

As the idea of using mouse movement analysis for identification originated from research conducted on keystroke dynamics, the initial approaches to using the characteristics of mouse movements analogously derived from the analysis of keyboard dynamics [33]. The problem faced by early works on identification based on mouse movements was the excessive amount of data necessary to recognize users—identification required as many as 2,000 actions [34] performed with a mouse, which rendered such solutions highly impractical [35]. The number of operations necessary for identification is directly related to the effectiveness of systems—the false acceptance rate for 500 actions analyzed is 7.78%; for 3,000 actions it is 0.37% [36]. The literature also describes methods based on non-typical movement indicators, such as angle-based metrics obtained point by point [32]. These types of measurement are unaffected by the computer environments in which users work. Their effectiveness barely exceeds one percent for both false-reject and false accept (which, nevertheless, remains above the upper thresholds dictated by European standards).

Recognition of users based on mouse movement parameters is occasionally combined with other methods of behavioral biometrics to increase the effectiveness of identification. This approach was applied, for instance, by Rose et al. [37], who combined recorded mouse movements with recorded eye movements via an eye-tracker. They found that combinations of these two signals enabled more accurate user identification. Although such systems might be applied in the future on a larger scale, their present development is hampered by the relative obscurity of equipment used to record eye movement.

4.3.2 Mouse Movement and Emotions

Users' emotions manifest in their behavior; this means that they could also be inferred from the way mice are used. Research on embodied cognition demonstrates that attitudes, cognitive states, and emotions manifest in the body's reactions that are involved in the development of feelings. Some scientists even advance the theory that understanding the world would be impossible without the activation of sensory areas. [38]. Neuroimaging studies also suggest that emotional processes could manifest in the movements of a hand that operates a mouse. A key role in the processes of controlled and goal-directed movement is played by the basal nuclei; they also form part of a system of connections between the cortex and the thalamus (the cortico-basal-ganglia-thalamo-cortical loop) [39, 40], covering the areas responsible for motor control and those related to the processing of emotions and decision-making [41]. The basal nuclei, the cortex, and the thalamus are also connected by the dopamine system. Dopamine is a neurotransmitter that participates in affective, motor, and cognitive processes by modulating the activity of neurons in the basal nuclei, limbic system, and cerebral cortex [42]. Functional disorders of the dopamine system manifest in motor and cognitive control deficits—in addition to emotional ones, such as Parkinson's disease [43]. The information presented above suggests that human emotions can be inferred from the way users move their mice.

The history of the analysis of mouse cursor movement dates back to Fitts's classic research on the ergonomics of airplane cockpits conducted in the 1950s and 1960s [44]. The studies were the basis for Fitts's law, according to which, in the case of rapid goal-oriented movements, compromises can be observed between their time and accuracy. Later, studies on cursor movements were conducted in which the effectiveness of different devices in text selection tasks was compared [45]. Correlations between cursor movements and emotions were investigated by Zimmermann et al. Although their studies demonstrated a relationship between the number of speed changes, the overall time of cursor movement, and emotional arousal, no correlation was identified between movement characteristics and the signs of emotions [13]. The results are inconclusive as Grimes et al. [46] demonstrated that exposure to negative stimuli increased the distance covered by mouse cursors. Other solutions documented in the literature involve the expansion of typical peripheral devices, such as mice and keyboards. By installing additional sensors, it is possible to non-intrusively measure a range of psychophysiological parameters that indicate emotions, such as temperature and electrodermal activity [47]. This type of sensor can be used in conjunction with measured parameters of mouse movement. Kaklauskas et al. [48] devised a system based on both mouse movement and additional parameters, including pressure applied by users on the casing and buttons of mice; electrodermal/electrogalvanic reaction of the skin; skin temperature; speed of mouse pointer movement; acceleration of mouse pointer movement; amplitude of hand shakiness; use of scroll wheels; and the frequency that both mouse buttons are clicked. The system allows users' emotions (anger, fear, sadness, disgust, happiness, and surprise) to be assessed, in addition to their productivity while working on computers. Similar systems have been proposed to measure the emotional states of students during exams [49].

4.3.3 Mouse Movements in Psychological Research

Mouse movement has been analyzed in many areas of psychology to better understand the mental processes that occur in the human brain when performing various types of tasks. Scientists have discovered that users often move their mouse not only after making decisions, but also during the process of making them. This results in "real-time motor traces of the mind" that can be recorded and analyzed [50]. Analyses of mouse cursor movements have been successfully applied in psychological research [51, 52] on decision-making, egoism, and implicit attitudes. The studies conducted by Maliszewski et al. in which subjects expressed their opinions on pictures painted by members of their own group and by members of a group of strangers, were designed to measure implicit attitudes. The results demonstrated that the parameters of mouse movement could serve as an indicator of favoritism of one's own group, which was impossible to discover by referring only to the subjects' explicit assessments. Psychologists also use mouse movement to study cognitive processes in social dilemma situations. Studies have analyzed games in which participants can either maximize their own payoffs (by choosing to withdraw from the game), ignoring the potential benefits or losses for their interaction partners; or cooperate to maximize collective payoffs [53]. Kieslich and Hilbig [54] analyzed mouse movements as an indicator of cognitive conflict in solving social dilemmas. They found that when making decisions to withdraw, mouse trajectories diverge more towards an unselected option than in decisions to cooperate.

Movement trajectory (involving, for example, uncertainty about marked judgments) may also be an indicator of difficulty in the decision-making process. In a study in which the task was to match terms (typically feminine and typically masculine) to faces depicted uniquely or ambiguously as masculine or feminine, it was shown that in the case of ambiguously masculine or feminine faces, the motion trajectory is similar to a non-matching face term because the connection between the term and the face image is ambiguous. [50]. Another indicator is the entropy of mouse movements. McKinstra et al. [55] demonstrated that when answering questions to which there was an unambiguous answer, entropy was lower; when answering ambiguous questions, mouse movements were more complex and entropy was higher. Their findings suggest that the nature of decision-making processes is continuous: that its piecemeal effects are sent, for example, to the motor system before the process is complete; and that while the mouse is still being moved, alternative choices remain able to 'attract' the moving mouse cursor. The parameters of this movement serve as indicators of the decision-making process behind such alternative choices. This theory is supported by research on the use of mouse movements to detect uncertainty. E-learning studies associate longer trajectories of mouse cursor movements with user indecision [56]. The analysis of mouse movements also allows researchers to assess how well the material learned is understood [57]. It has been observed that indecision, or cognitive conflict, can manifest in hand movements [58]. This has been demonstrated in studies that do not pertain directly to mouse movements, but those focused on arm movements recorded with the use of Nintendo Wii controllers.

In the fields of human–computer interaction and user experience, mouse movement analysis facilitates understanding of users' thought processes while working on computers. This, in turn, allows the identification of website sections that are easy or difficult to use, and those that are of most interest to users. In web design usability tests, subjects who are unconfident in performing their tasks move the mouse more slowly, while those who are confident operate it faster [59, 60]. It is also possible to identify correlations of which users may not be aware, as is the case with research on illicit stereotypes. Mueller and Lockerd [61] were able to predict, for example, which options were specific users' second choices by analyzing the trajectories of their cursors over the options displayed on screens.

4.3.4 Mouse Movement in Survey Research

Mouse movement information is applied in user identification, user emotion recognition, and psychological research—but also serves as a source of additional data in computer-based surveys. One challenge of surveys is the subjective difficulty of the questions, which causes misunderstandings and produces incorrect answers [62]. Misunderstandings can result from, among other things, unfamiliarity with the vocabulary used in the questions and answers, poorly worded questions, or factors related to the respondents, such as their inability to relate the issues highlighted in the questions to their own experiences. For that reason, the content of survey questions is analyzed meticulously to ensure that all questions are as understandable as possible to the respondents. Traditionally, in the case of in-person surveys, such problems were handled by interviewers; presently, however, when most of the surveys are conducted online, polling organizations seek other indicators to assist in identification of respondents' problems understanding the questions, including reaction times [63] and mouse movement parameters [64]. While response times are easy to record and can serve as a reliable difficulty indicator [65], mouse movement data enables opportunities to analyze the survey response process more deeply. It has been demonstrated, for example, that long mouse movements (those that exceed the average for the studied group by two standard deviations) are a reliable indicator of poor data quality for the questions in which they occurred [66]. Subsequent studies, which sought new indicators reaching beyond the distance covered by cursors, involved testing indicators used in the field of human–computer interaction studies [64], such as hovering and regressions. To identify difficult survey questions, the following indicators were considered useful: hovering the cursor over the question text for more than two seconds; hovering the cursor over the answer selection spot (for instance, a checkbox or radio button); and moving the cursor between the text of the question, the answers, and blank spaces on the screen. The authors emphasize, however, that the correlations observed might only be true for specific types of question.

References

1. Yamada, H.: A historical study of typewriters and typing methods, from the position of planning Japanese parallels (1980)
2. Yasuoka, K., Yasuoka, M.: On the prehistory of QWERTY. Zinbun. **42**, 161–174 (2011)
3. Kumar, B.P., Reddy, E.S.: An efficient security model for password generation and time complexity analysis for cracking the password (2020). https://doi.org/10.18280/ijsse.100517
4. Monrose, F., Rubin, A.D.: Keystroke dynamics as a biometric for authentication. Future Gener. Comput. Syst. **16**, 351–359 (2000)
5. Bergadano, F., Gunetti, D., Picardi, C.: Identity verification through dynamic keystroke analysis. Intell. Data Anal. **7**, 469–496 (2003)
6. Joyce, R., Gupta, G.: Identity authentication based on keystroke latencies. Commun. ACM. **33**, 168–176 (1990)
7. Dowland, P.S., Furnell, S.M.: A long-term trial of keystroke profiling using digraph, trigraph and keyword latencies (2004). https://doi.org/10.1007/1-4020-8143-x_18
8. Gunetti, D., Picardi, C.: Keystroke analysis of free text (2005). https://doi.org/10.1145/1085126.1085129
9. Gunetti, D., Picardi, C., Ruffo, G.: Keystroke analysis of different languages: a case study. In: Advances in Intelligent Data Analysis VI, pp. 133–144. Springer Berlin Heidelberg (2005).
10. Kopacz, A., Biele, C., Zdrodowska, A.: Development and validation of a shortened language-specific version of the UNRAVEL placekeeping ability performance measuring tool. Adv. Cogn. Psychol. **15**, 256–264 (2019)
11. Brown, M., Rogers, S.J.: User identification via keystroke characteristics of typed names using neural networks. Int. J. Man. Mach. Stud. **39**, 999–1014 (1993)
12. Sheng, Y., Phoha, V.V., Rovnyak, S.M.: A parallel decision tree-based method for user authentication based on keystroke patterns. IEEE Trans. Syst. Man Cybern. B Cybern. **35**, 826–833 (2005)

13. Zimmermann, P., Guttormsen, S., Danuser, B., Gomez, P.: Affective computing—a rationale for measuring mood with mouse and keyboard. Int. J. Occup. Saf. Ergon. **9**, 539–551 (2003)
14. Epp, C., Lippold, M., Mandryk, R.L.: Identifying emotional states using keystroke dynamics. In: Proceedings of the SIGCHI Conference on Human Factors in Computing Systems, pp. 715–724. Association for Computing Machinery, New York, NY, USA (2011)
15. Karnan, M., Akila, M.: Identity authentication based on keystroke dynamics using genetic algorithm and particle Swarm Optimization. In: 2009 2nd IEEE International Conference on Computer Science and Information Technology, pp. 203–207 (2009)
16. Kay, R., Lauricella, S.: Assessing laptop use in higher education: the laptop use scale. J. Comput. High. Educ. **28**, 18–44 (2016)
17. Barak, M., Lipson, A., Lerman, S.: Wireless laptops as means for promoting active learning in large lecture halls. Int. J. Inf. Commun. Technol. Educ. **38**, 245–263 (2006)
18. Patterson, R.W., Patterson, R.M.: Computers and productivity: evidence from laptop use in the college classroom (2017). https://doi.org/10.1016/j.econedurev.2017.02.004
19. Yamamoto, K.: Banning laptops in the classroom: is it worth the hassles? J. Legal Educ. **57**, 477–520 (2007)
20. Kiewra, K.A.: A review of note-taking: the encoding-storage paradigm and beyond. Educ. Psychol. Rev. **1**, 147–172 (1989)
21. Intons-Peterson, M.J., Fournier, J.: External and internal memory aids: when and how often do we use them? J. Exp. Psychol. Gen. **115**, 267–280 (1986)
22. Van Meter, P., Yokoi, L., Pressley, M.: College students' theory of note-taking derived from their perceptions of note-taking. J. Educ. Psychol. **86**, 323–338 (1994)
23. Igo, L.B., Bruning, R., McCrudden, M.T.: Exploring differences in students' copy-and-paste decision making and processing: a mixed-methods study. J. Educ. Psychol. **97**, 103–116 (2005)
24. Mueller, P.A., Oppenheimer, D.M.: The pen is mightier than the keyboard: advantages of longhand over laptop note taking. Psychol. Sci. **25**, 1159–1168 (2014)
25. Craik, F.I.M., Tulving, E.: Depth of processing and the retention of words in episodic memory (1975). https://doi.org/10.1037/0096-3445.104.3.268
26. Umejima, K., Ibaraki, T., Yamazaki, T., Sakai, K.L.: Paper notebooks versus mobile devices: Brain activation differences during memory retrieval. Front. Behav. Neurosci. **15**, 34 (2021)
27. Planton, S., Jucla, M., Roux, F.-E., Démonet, J.-F.: The "handwriting brain": a meta-analysis of neuroimaging studies of motor versus orthographic processes. Cortex **49**, 2772–2787 (2013)
28. Higashiyama, Y., Takeda, K., Someya, Y., Kuroiwa, Y., Tanaka, F.: The neural basis of typewriting: a functional MRI study (2015). https://doi.org/10.1371/journal.pone.0134131
29. Longcamp, M., Hlushchuk, Y., Hari, R.: What differs in visual recognition of handwritten versus printed letters? An fMRI study. Hum. Brain Mapp. **32**, 1250–1259 (2011)
30. Freyd, J.J.: The mental representation of movement when static stimuli are viewed. Percept. Psychophys. **33**, 575–581 (1983)
31. Engelbart, D.C., English, W.K.: A research center for augmenting human intellect. In: Proceedings of the December 9–11, 1968, fall joint computer conference, part I, pp. 395–410. Association for Computing Machinery, New York, NY, USA (1968)
32. Zheng, N., Paloski, A., Wang, H.: An efficient user verification system using angle-based mouse movement biometrics. ACM Trans. Inf. Syst. Secur. **18**, 1–27 (2016)
33. Ahmed, A.A.E., Traore, I.: Anomaly intrusion detection based on biometrics. In: Proceedings from the Sixth Annual IEEE SMC Information Assurance Workshop, pp. 452–453 (2005)
34. Nakkabi, Y., Traore, I., Ahmed, A.A.E.: Improving mouse dynamics biometric performance using variance reduction via extractors with separate features. IEEE Trans. Syst. Man Cybern. A Syst Human. **40**, 1345–1353 (2010)
35. Ahmed, A.A.E., Traore, I.: A new biometric technology based on mouse dynamics. IEEE Trans. Depend. Secure Comput. **4**, 165–179 (2007)
36. Shen, C., Cai, Z., Guan, X.: Continuous authentication for mouse dynamics: A pattern-growth approach. In: IEEE/IFIP International Conference on Dependable Systems and Networks (DSN 2012), pp. 1–12 (2012)

37. Rose, J., Liu, Y., Awad, A.: Biometric authentication using mouse and eye movement data (2017). https://doi.org/10.1109/spw.2017.18
38. Barsalou, L.W., Niedenthal, P.M., Barbey, A.K., Ruppert, J.A.: Social embodiment. In: Ross, B.H. (ed.) The Psychology of Learning and Motivation: Advances in Research and Theory, pp. 43–92. Elsevier Science, New York, NY, US (2003)
39. Parent, A., Hazrati, L.N.: Functional anatomy of the basal ganglia. I. The cortico-basal ganglia-thalamo-cortical loop. Brain Res. Brain Res. Rev. **20**, 91–127 (1995)
40. Graybiel, A.M., Aosaki, T., Flaherty, A.W., Kimura, M.: The basal ganglia and adaptive motor control. Science **265**, 1826–1831 (1994)
41. Graybiel, A.M.: The basal ganglia. Curr. Biol. **10**, R509–R511 (2000)
42. Björklund, A., Dunnett, S.B.: Dopamine neuron systems in the brain: an update. Trends Neurosci. **30**, 194–202 (2007)
43. Rabey, J.M.: Neurobehavioral disorders in Parkinson's disease. Handb. Clin. Neurol. **83**, 435–455 (2007)
44. Fitts, P.M.: Information capacity of the human motor system in controlling the amplitude of movement (1954)
45. Card, S.K., English, W.K., Burr, B.J.: Evaluation of mouse, rate-controlled isometric joystick, step keys, and text keys for text selection on a CRT. Ergonomics **21**, 601–613 (1978)
46. Grimes, M., Jenkins, J., Valacich, J.: Exploring the effect of arousal and valence on mouse interaction. In: ICIS 2013 Proceedings (2013)
47. Sung, K.Y.: A suggestion to improve user-friendliness based on monitoring computer user's emotions (2017). https://doi.org/10.1007/978-3-319-58637-3_10
48. Kaklauskas, A., Zavadskas, E.K., Seniut, M., Krutinis, M., Dzemyda, G., Ivanikovas, S., Stankevic, V., imkevicius, C., Jaruevicius, A.: Web-based biometric mouse decision support system for users emotional and labour productivity analysis (2008). https://doi.org/10.22260/isarc2008/0013
49. Kaklauskas, A., Krutinis, M., Seniut, M.: Biometric mouse intelligent system for student's emotional and examination process analysis (2009). https://doi.org/10.1109/icalt.2009.130
50. Freeman, J.B., Ambady, N.: Motions of the hand expose the partial and parallel activation of stereotypes. Psychol. Sci. **20**, 1183–1188 (2009)
51. Maliszewski, N., Wojciechowski, Ł, Suszek, H.: Spontaneous movements of a computer mouse reveal egoism and in-group favoritism. Front. Psychol. **8**, 13 (2017)
52. Yamauchi, T., Leontyev, A., Wolfe, M.: Choice reaching trajectory analysis as essential behavioral measures for psychological science. Insight Psychol. **1**, 1 (2017)
53. Dawes, R.M.: Social dilemmas. Annu. Rev. Psychol. **31**, 169–193 (1980)
54. Kieslich, P.J., Hilbig, B.E.: Cognitive conflict in social dilemmas: an analysis of response dynamics. Judgm. Decis. Mak. **9**, 510 (2014)
55. McKinstry, C., Dale, R., Spivey, M.J.: Action dynamics reveal parallel competition in decision making. Psychol. Sci. **19**, 22–24 (2008)
56. Zushi, M., Miyazaki, Y., Norizuki, K.: Web application for recording learners' mouse trajectories and retrieving their study logs for data analysis. Knowl. Manag. E-Learn. Int. J. **4**, 37–50 (2012)
57. Zushi, M., Miyazaki, Y., Norizuki, K.: Web application for measuring learners' knowledge of english syntax: analyzing the relationship between mouse trajectories and learners' understanding (2014). https://doi.org/10.3366/ijhac.2014.0107
58. Duran, N.D., Dale, R., McNamara, D.S.: The action dynamics of overcoming the truth. Psychon. Bull. Rev. **17**, 486–491 (2010)
59. Arroyo, E., Selker, T., Wei, W.: Usability tool for analysis of web designs using mouse tracks. In: CHI '06 Extended Abstracts on Human Factors in Computing Systems, pp. 484–489. Association for Computing Machinery, New York, NY, USA (2006)
60. Huang, J., White, R.W., Dumais, S.: No clicks, no problem: using cursor movements to understand and improve search. In: Proceedings of the SIGCHI Conference on Human Factors in Computing Systems, pp. 1225–1234. Association for Computing Machinery, New York, NY, USA (2011)

61. Mueller, F., Lockerd, A.: Cheese: tracking mouse movement activity on websites, a tool for user modeling. In: CHI '01 Extended Abstracts on Human Factors in Computing Systems, pp. 279–280. Association for Computing Machinery, New York, NY, USA (2001)
62. Ehlen, P., Schober, M.F., Conrad, F.G.: Modeling speech disfluency to predict conceptual misalignment in speech survey interfaces. Discourse Process. **44**, 245–265 (2007)
63. Conrad, F.G., Schober, M.F., Coiner, T.: Bringing features of human dialogue to web surveys. Appl. Cogn. Psychol. **21**, 165–187 (2007)
64. Horwitz, R., Kreuter, F., Conrad, F.: Using mouse movements to predict web survey response difficulty. Soc. Sci. Comput. Rev. **35**, 388–405 (2017)
65. Yan, T., Tourangeau, R.: Fast times and easy questions: the effects of age, experience and question complexity on web survey response times. Appl. Cogn. Psychol. **22**, 51–68 (2008)
66. Stieger, S., Reips, U.-D.: What are participants doing while filling in an online questionnaire: a paradata collection tool and an empirical study. Comput. Hum. Behav. **26**, 1488–1495 (2010)

Chapter 5
Leg and Foot Movement

5.1 Introduction

Systems used in human–computer interaction are typically optimized for manual operation and accommodate the full range of hand movement and capabilities. Such systems may, however, also be effected through other means: interfaces operated with the feet and legs may be used not only as an alternative for those who are unable to use their hands, but also as fully operational interfaces with their own advantages and applications. The first mentions of foot-operated interfaces that appeared in the literature of the subject can be attributed to Douglas Engelbart, inventor of the computer mouse. Since then, research in human–computer interaction and other disciplines has given rise to a large number of foot-operated interfaces, which are used, for example, in mobile solutions [1] and interactive floors.

In pursuance of efficient foot interfaces, it is useful for developers to know how the feet are built, and to understand both the limitations imposed and the opportunities provided by their anatomy—which, coupled with the role it plays in psychological functioning, dictates how the feet can be involved in interactions.

5.2 Anatomy of the Foot

As an insightful observer and a conscientious researcher, Leonardo da Vinci asserted that "the human foot is a masterpiece of engineering and a work of art." Feet are found only in primates; and only those of humans allow upright, bipedal movement, owing to their unique anatomy. Changes to human lifestyles, living conditions, and work necessitated the evolution of the feet. Humans continue to benefit from the results, as it is the upright position that has influenced the shape of the foot as we know it.

C. Biele, *Human Movements in Human-Computer Interaction (HCI)*,
Studies in Computational Intelligence 996,
https://doi.org/10.1007/978-3-030-90004-5_5

Human feet have discarded their opposable, grasping big toes, which once facilitated brachiation; simian feet are built like hands, and are thus capable of grasping. The reduction in size of the opposable big toe in favor of elongation of the first metatarsal bone allowed humans to stand and walk comfortably and steadily. The big toe acts as one of three points of contact with the ground that enable this type of movement. The upright position, which emerged in tandem with evolutionary changes in the foot's anatomy, offer humans an evolutionary advantage: it makes it easier to collect food from short trees and bushes, as well as carrying larger amounts of it. The upright position also protects the skin from exposure to direct sunlight and eases both the use of tools in the hands and the burden of carrying children across large distances. Moving on two legs consumes less energy when walking slowly and ensures enhanced safety as the field of vision expands, and the human body is perceived as larger (and more threatening) by potential predators [1].

The human foot comprises 26 bones, more than 30 joints, and more than 100 ligaments that enable it to perform its functions: walking, maintaining balance, and absorbing shocks. From below, a well-formed foot is triangular: it is broad at the toes and narrow at the heel. The foot makes contact with surfaces using the calcaneal tuberosity and the heads of the first and fifth metatarsals: these points act as the vertices of the aforementioned triangle in which each of the edges forms one of the foot arches (two longitudinal and one transverse). The bones of the foot connect with tendons via ligaments, which maintains them in the form of an arch, while allowing easy movement—resulting in overall high elasticity. During movement, such as jogging, the foot arches are subject to significant strain, with energy stored in them for the purpose of propulsion. The triangular anatomy enables not only shock absorption while walking, running, and performing other actions, but also the adoption of stable standing positions, as well as the handling of heavy loads [2]. Standing and maintaining balance (so-called "static foot" posture) and providing propulsion to the body when moving ("dynamic foot") are the primary functions of human feet [3, 4].

From the perspective of human–computer interaction, the absence of the grasping function in humans facilitates easy adaption of interfaces designed for manual operation, such as mice and joysticks, for use with the feet. The recognizable shape of feet also makes it easier to identify images of them in computer vision systems [5].

5.3 Foot and Leg Movement from a Psychological Perspective

In the course of evolution, the lower limbs have become the primary means of locomotion, as well as of escape from threats; consequently, the limbic system, which is responsible for processing emotions, also plays a major role in managing motor reactions. Evidence for reactions of this type can be sourced from studies [6] demonstrating that individuals experiencing anxiety during foreign language exams tend to cross their legs and change positions more frequently.

Scientific and anecdotal evidence suggests that foot activity reflects psychological processes; the feet and legs are valuable sources of involuntary non-verbal signals. In social interactions, humans have a tendency to focus on each other's faces, while the activity of the feet and legs is subject to less conscious control. This means that the lower limbs are of uncommon value as indicators of emotional condition, particularly when verbal communication stands in opposition to behavioral signals—a phenomenon known as "non-verbal leakage". Assumptions of this kind have their roots in the observations of Charles Darwin himself, who defined three primary principles that shape emotional expressions: the principle of serviceable habits, the principle of antithesis, and the principle of direct action. The first pertains to behaviors aimed at survival that are performed voluntarily in response to dangerous environmental stimuli (escaping a predator is a prime example); therefore, subconsciously, such actions are connected with mental states, like fear. As a result of this connection, a mental state begins to trigger a preparatory reaction—for instance, a suitable position of the feet, or muscle tension [7]. Pursuant to the principle of antithesis, opposing emotions trigger opposing physical actions, such as approach and avoidance, extension of the arm towards a positive stimulus, or its retraction from a stimulus that causes fear. The principle of direct action states that excitation of the nervous system directly affects the whole body and its reactions. In the case of foot arrangement, the first two of Darwin's concepts apply: changes in the arrangement of feet are associated with preparing the body for action; while their direction (toward/away from the stimulus) is related to the actions' meanings (positive/negative). Similar conclusions can be drawn from the position of the legs and whether or not they are crossed. Positive leg crossing involves arranging the legs toward one's interaction partner, which exposes the physiologically vulnerable area of the inner thigh—where the femoral artery is located; when the legs are crossed away from one's interaction partner, this area remains hidden. Arranging the legs in a position that exposes their internal sections is a subconscious way of displaying trust and positive attitudes toward others. Modern research has been conducted on this subject by Paul Ekman, famous for his studies on emotions. Ekman's work on non-verbal communication is largely based upon Darwin's concepts [8].

An individual's mental state may be revealed not only in how they move, but even through the arrangement of their feet and legs [9]. Symmetry of lower limb arrangement can act as a measure of relaxation: when high degrees of stress are experienced, the feet rest flat on the ground, often touching; when stress levels are low, one foot is frequently lifted.

Human behavior, including movement of the legs and feet, is also affected on the subconscious level by the actions of other interaction participants: this is the so-called chameleon effect, which is used to describe individuals' tendency to subconsciously emulate the behavior of others in their presence. In a study by Chartrand and Bargh [10], participants tapped their feet more frequently during the experiment if their companions were doing the same.

The arrangement of the legs and feet is also related to individuals' perceptions of those with whom they interact. Studies [11] have demonstrated that physicians who enjoy warm relations with their patients more frequently sit without crossing their legs, their bodies facing the patients.

Behaviors related to arm and leg movement are rooted in cultural backgrounds. In parts of Asia, pointing with one's feet or displaying their soles might be considered insulting. It transfers to the area of human computer interaction as in studies on touch interfaces [12], touching on-body targets on the lower limbs proved to be less socially acceptable than touching those on the upper limbs.

Potential relationships between emotions and feet have also been explored in alternative medicine. Erkek and Atkas [13] studied the effectiveness of foot massage in reducing anxiety levels during childbirth; women at various stages of the process were surveyed using the STAI TX-1 questionnaire—a Likert-type scale developed by Spielberger et al. in 1964 to measure levels of anxiety in healthy and sick individuals [14]. Foot massage relieves cramps and relaxes the muscles, which, in turn, reduces the transmission of pain stimuli to the brain, resulting in lower anxiety levels [15]. This is crucial during childbirth, as reduced anxiety makes it easier for women to withstand labor pains. The effectiveness of foot massage in pain reduction has similarly been demonstrated in patients who have undergone cardiac surgery [16]. It is also conducive to improvements in mood and reductions in anxiety among cancer patients [17].

5.4 Leg and Foot Interaction

The body of research on computer interactions involving the feet has grown for several reasons. The legs are a viable alternative for users who are unable to use their hands –due to medical conditions, for example. In some cases, it may be easier to reach certain elements using the feet than the hands, and certain types of interaction in virtual environments, such as locomotion, are designed with this in mind [18]. The legs and feet also constitute additional sources of signals in interactions based on other modalities [19].

Body position is a factor of significant influence on the manner of interaction with computer systems involving the legs and feet. It allows researchers to distinguish three types of interaction, each with its own set of specific limitations. In the case of the sitting position, which is the most frequent style of interaction with traditional computers, the range of motion for feet or legs is limited due to the space available under the desk; although, solutions of this type have the advantage of allowing actions, such as the lifting or depressing of pedals, to be performed with both feet. Their downside, however, is that the structure of desks prevents visual feedback, making solutions that require users to see areas of interaction impracticable [5]. Adopting a sitting position also results in relatively swift muscle fatigue when performing actions using the legs; conversely, maintaining balance could be problematic in standing positions. These difficulties reduce the number of possible actions that can be used in interactions of the type implemented in public displays (such as games) to take advantage of users' wide range of motions. The third type of interaction involving the feet occurs when walking or jogging, in which the range of motions is even greater; limiting cognitive and attention resources constitute a risk, however, as some are involved in the walking process itself. Moreover, performing actions using the legs can prove challenging, as movement remains their primary task. Movement itself serves as the input signal (the subject of movement in the environment is discussed in more detail in Chap. 8). This type of interaction is often utilized in artistic installations.

5.4.1 Leg and Foot Interfaces

Interfaces designed for foot operation are highly diverse, ranging from small devices similar to computer mice that are operated using the feet to large, interactive floors that comprise entire rooms. There are three methods of data collection on foot movement: indirectly, usually through the use of devices, such as mice, switches, and pedals; with the aid of sensors placed on the feet—for instance, augmented shoes; and with the aid of sensors installed in the environment, such as lasers, infrared cameras, and augmented floors.

5.4.1.1 Indirect Interfaces

Pedals are the most commonly used indirect sensors. Studies of their implementations were conducted prior, even, to the advent of the computer, due to their role in operating machinery, in driving cars, and in playing musical instruments [20]. Pedals continue to be used widely in music, including in pianos, and in guitar and synthesizer effects. Pedals can take the form of simple switches (momentary or latching), or enable stepless adjustment of parameter values. The first foot-operated solutions in this context were proposed by Douglas Engelbart, inventor of the mouse [21]. Since then, the usefulness of pedals in human–computer interaction has been studied in myriad areas, including in operating 3D modeling applications [22], multi modal text entry [23], radiological interventions [24], map navigation [25], and the use of piano keyboards as controllers [26]. Foot mice [22], foot joysticks [27], foot trackballs [28], and pressure-measuring boards—which have gained commercial recognition in the form of the Balance Board attachment for the Nintendo Wii—have also been studied. The usefulness of such interfaces has been examined, for example, with respect to maps [29] and the navigation of 3D environments [30].

When discussing interfaces with indirect foot controls, it is worth mentioning interfaces that augment movement in V that allow foot movements to be translated into movement within virtual environments. The most common solutions are step-in-place devices [31, 32], foot platforms, treadmills, and pedaling devices.

5.4.1.2 Direct Interfaces

Direct sensors utilize measurement methods that record data received from devices installed in shoes. Pressure sensors inside the soles provide data on the weight distribution on the foot's surface, while flexion sensors allow the degree of sole flexion to be examined. Typically, the data is recorded in the background. The usefulness of direct recording of signals from the feet to control prosthetic hands was tested by Carozza et al. [33]. Systems that utilize pressure sensors installed in soles have been evaluated as easier to use and operate than electromyographic signal-based interfaces. Patterns of pressure on sole points, sole flexion, and movement parameters like acceleration are sufficiently specific to each individual to enable their identification [34].

Studies pertaining to control of human–computer interaction using the feet have also been conducted. Shoes equipped with additional sensors have been used as replacements for conventional peripherals, such as keyboards [35] or mice (see: Shoe-Mouse, for instance [36]).

Data collected by sensors installed in footwear are also frequently used in artistic performances [37, 38] and in VR environments. Matthies et al. [39] developed a prototypical system that enables users to move around in CAVE-type VR installations by walking on the spot.

Sensor-equipped shoes have also found applications in medicine. Morley et al. [40] used temperature, humidity, and pressure sensors to monitor conditions inside the shoes of diabetes patients. When combined with analysis of foot movements and sole pressure, optical sensors can be used to enable individuals with hand disabilities to write [41]. Users are able to use their foot movements to "draw" letter shapes on the ground, acting as a substitute for handwriting. Tao et al. [42] have outlined a fall detection system, which might be used to monitor the condition of the elderly. The system utilizes force-sensing resistors, whose signals are analyzed in real time, enabling the detection not only of fall events themselves, but also their direction.

5.4.1.3 Environmental Interfaces

Another category of solution comprises those based on sensors distributed in the environment. In such cases, little to no hardware is required to be worn by users—a considerable advantage. Solutions that require only slight modifications to users' footwear are commonly based on infrared technology. Cameras are installed in the environment, with markers reflecting infrared light attached to individuals' shoes or legs. For such systems to be effective, multiple cameras must be used to ensure that the sensors remain within their field of view. Examples of infrared-based technological solutions include the Fantastic Phantom Slipper [43] (a pair of slippers used for interacting with a virtual environment), and a system capable of determining the general direction of users' vision from the position of their feet (the system's accuracy for fifteen-degree angles was 60%) [44].

Regular cameras used to detect the position of feet, when coupled with image processing algorithms, can be incorporated in augmented reality solutions, such as

AR-soccer, which was developed in the early years of the twenty-first century for personal digital assistants [45]. The application detected the shape of users' feet, and their contact with a virtual ball displayed on the screen; thus, making it possible to "kick" the ball. Presently, smartphones' popularity, computing capacity, and the high quality of their cameras have resulted in applications of this type becoming commonplace (for instance, Champions AR Soccer [46]).

Interactive floors are also used in human–computer interaction. They are commonly available in two versions: sensor- and camera-based. Sensor-based interactive floors have primarily been used in art and dance.

The Litefoot system [47] developed at the University of Limerick utilized a panel with a surface area of nearly two square meters with close to 2,000 proximity sensors installed to determine foot positions, as well as an accelerometer to measure the force of footfalls. Signals received from the panel were used to create visualizations and sound effects played during performances of Irish dance.

Magic Carpet [48] was a 5×10 m carpet with an embedded grid of piezoelectric wires, which generate electricity when bent. The distribution of wires allowed users to be located as they walked on the carpet. Moreover, the system was equipped with Doppler-based motion detectors to track upper body and arm movements.

Interactive floors are not limited to the rectangular, however. Richardson et al. [49] proposed Z-Tiles—a sensing floor that can be arranged freely like a jigsaw puzzle from irregularly-shaped pieces made of hexagonal tiles equipped with pressure sensors. This enables construction of sensing floors of any shape, which widens the variety of their potential applications.

Some systems allow multiple users to interact with them concurrently. A floor of this type was designed as part of the Virtual Space project [50]. It utilizes sensors that allow simultaneous spatial interaction for multiple users, and was tested as a method of controlling a Pong-inspired video game displayed on a vertical screen.

Another type of sensing floor utilizes cameras that track the positions of users' feet via computer vision algorithms. Multitoe [6], a high-resolution floor that enables highly accurate data input using the feet, is one such solution discussed in the literature—the rate of error in its best performing model is a mere 3%. One example of a camera-based solution is the iGameFloor [51]—a system that utilizes webcams installed below a transparent floor to track users' positions. Content is projected onto the floor's surface from below. Owing to its surface area of 12m^2 and its ability to simultaneously track more than 40 points, the system facilitates multiuser interaction effectively.

The capabilities of sensing floors have also been used commercially. Sony released a dancing game titled Dance Dance Revolution (DDR) for its Playstation console sold with a sensing mat that detected players' footfalls. The game and mat have also proven useful in scientific research, likely due to their affordability and portability. DDR has been used in studies on the functioning of the elderly [52] and on creativity [53], as well as those on obesity in children and promoting fitness [54, 55].

5.5 Conclusions

As you can see, feet and legs can find applications in the area of human-technology interaction. Currently, they are mainly used in this context as an interaction tool either replacing typical control devices or as completely new HCI methods. An interesting area that has not been explored so far is the question of reading e.g. emotional state or arousal on the basis of computer analysis of the user's leg movement. Current solutions including motion analysis include also legs, but they are treated as a part of whole body motion analysis i.e. [56].

References

1. Weaver, T.D., Klein, R.G.: The evolution of human walking. Human Walking, pp. 23–32 (2006)
2. Dawe, E.J.C., Davis, J.: (vi) Anatomy and biomechanics of the foot and ankle (2011). https://doi.org/10.1016/j.mporth.2011.02.004
3. Bramble, D.M., Lieberman, D.E.: Endurance running and the evolution of Homo. Nature **432**, 345–352 (2004)
4. Ridola, C., Palma, A.: Functional anatomy and imaging of the foot. Ital. J. Anat. Embryol. **106**, 85–98 (2001)
5. Augsten, T., Kaefer, K., Meusel, R., Fetzer, C., Kanitz, D., Stoff, T., Becker, T., Holz, C., Baudisch, P.: Multitoe: high-precision interaction with back-projected floors based on high-resolution multi-touch input. In: Proceedings of the 23nd annual ACM Symposium on User Interface Software and Technology, pp. 209–218. Association for Computing Machinery, New York, NY, USA (2010)
6. Gregersen, T.S.: Nonverbal cues: clues to the detection of foreign language anxiety. Foreign Lang. Ann. **38**, 388–400 (2005)
7. Darwin, C.: The expression of the emotions in man and animals (2015).
8. Ekman, P., Friesen, W.V.: Nonverbal leakage and clues to deception. Psychiatry **32**, 88–106 (1969)
9. Mehrabian, A.: Some referents and measures of nonverbal behavior. Behav. Res. Method. Instrum. **1**, 203–207 (1968)
10. Chartrand, T.L., Bargh, J.A.: The chameleon effect: the perception–behavior link and social interaction. J. Pers. Soc. Psychol. **76**, 893–910
11. Harrigan, J.A., Oxman, T.E., Rosenthal, R.: Rapport expressed through nonverbal behavior. J. Nonverbal Behav. **9**, 95–110 (1985)
12. Wagner, J., Nancel, M., Gustafson, S.G., Huot, S., Mackay, W.E.: Body-centric design space for multi-surface interaction. In: Proceedings of the SIGCHI Conference on Human Factors in Computing Systems, pp. 1299–1308. Association for Computing Machinery, New York, NY, USA (2013)
13. Yılar Erkek, Z., Aktas, S.: The effect of foot reflexology on the anxiety levels of women in labor. J. Altern. Complement. Med. **24**, 352–360 (2018)
14. Spielberger, C.D., Gorsuch, R.L., Lushene, R.E.: STAI manual for the state-trait anxiety inventory ("self-evaluation questionnaire"). Consult. Psychol. (1970)
15. Hoeger Bement, M., Weyer, A., Keller, M., Harkins, A.L., Hunter, S.K.: Anxiety and stress can predict pain perception following a cognitive stress. Physiol. Behav. **101**, 87–92 (2010)
16. Alameri, R., Dean, G., Castner, J., Volpe, E., Elghoneimy, Y., Jungquist, C.: Efficacy of precise foot massage therapy on pain and anxiety following cardiac surgery: pilot study. Pain Manag. Nurs. **21**, 314–322 (2020)

17. Noh, G.O., Park, K.S.: Effects of aroma self-foot reflexology on peripheral neuropathy, peripheral skin temperature, anxiety, and depression in gynaecologic cancer patients undergoing chemotherapy: a randomised controlled trial. Eur. J. Oncol. Nurs. **42**, 82–89 (2019)
18. Drossis, G., Grammenos, D., Bouhli, M., Adami, I., Stephanidis, C.: Comparative evaluation among diverse interaction techniques in three dimensional environments. In: Distributed Ambient, and Pervasive Interactions, pp. 3 12. Springer Berlin Heidelberg (2013)
19. Göbel, F., Klamka, K., Siegel, A., Vogt, S., Stellmach, S., Dachselt, R.: Gaze-supported foot interaction in zoomable information spaces (2013). https://doi.org/10.1145/2468356.2479610
20. Rosenblum, S.P.: Pedaling the piano: a brief survey from the eighteenth century to the present. Perform. Pract Rev. (1993)
21. English, W.K., Engelbart, D.C., Berman, M.L.: Display-selection techniques for text manipulation. IEEE Transactions on Human Factors in Electronics. HFE-8, pp. 5–15 (1967)
22. Balakrishnan, R., Fitzmaurice, G., Kurtenbach, G., Singh, K.: Exploring interactive curve and surface manipulation using a bend and twist sensitive input strip. In: Proceedings of the 1999 Symposium On Interactive 3D Graphics, pp. 111–118. Association for Computing Machinery, New York, NY, USA (1999)
23. Dearman, D., Karlson, A., Meyers, B., Bederson, B.: Multi-modal text entry and selection on a mobile device. In: Proceedings of Graphics Interface 2010, pp. 19–26 (2010)
24. Hatscher, B., Luz, M., Hansen, C.: Foot interaction concepts to support radiological interventions. i-com. **17**, 3–13 (2018)
25. Alexander, J., Han, T., Judd, W., Irani, P., Subramanian, S.: Putting your best foot forward: investigating real-world mappings for foot-based gestures. In: Proceedings of the SIGCHI Conference on Human Factors in Computing Systems, pp. 1229–1238. Association for Computing Machinery, New York, NY, USA (2012)
26. Mohamed, F., Fels, S.: LMNKui: overlaying computer controls on a piano controller keyboard. In: CHI '02 Extended Abstracts on Human Factors in Computing Systems, pp. 638–639. Association for Computing Machinery, New York, NY, USA (2002)
27. Garcia, F.P., Vu, K.P.L.: Effectiveness of hand- and foot-operated secondary input devices for word-processing tasks before and after training (2011). https://doi.org/10.1016/j.chb.2010.08.006
28. Pakkanen, T., Raisamo, R.: Appropriateness of foot interaction for non-accurate spatial tasks. In: CHI '04 Extended Abstracts on Human Factors in Computing Systems, pp. 1123–1126. Association for Computing Machinery, New York, NY, USA (2004)
29. Schöning, J., Daiber, F., Krüger, A., Rohs, M.: Using hands and feet to navigate and manipulate spatial data. In: CHI '09 Extended Abstracts on Human Factors in Computing Systems, pp. 4663–4668. Association for Computing Machinery, New York, NY, USA (2009)
30. Filho, E.X. de L., de Lima Filho, E.X., Nunes, M.B., Comba, J., Nedel, L.: Why not with the foot? (2011). https://doi.org/10.1109/sbgames.2011.33
31. Templeman, J.N., Denbrook, P.S., Sibert, L.E.: Virtual locomotion: walking in place through virtual environments. Presence. **8**, 598–617 (1999)
32. Bouguila, L., Evequoz, F., Courant, M., Hirsbrunner, B.: Walking-pad: a step-in-place locomotion interface for virtual environments. In: Proceedings of the 6th International Conference on Multimodal Interfaces, pp. 77–81. Association for Computing Machinery, New York, NY, USA (2004)
33. Carrozza, M.C., Persichetti, A., Laschi, C., Vecchi, F., Lazzarini, R., Vacalebri, P., Dario, P.: A wearable biomechatronic interface for controlling robots with voluntary foot movements. IEEE/ASME Trans. Mechatron. **12**, 1–11 (2007)
34. Huang, B., Chen, M., Ye, W., Xu, Y.: Intelligent shoes for human identification. In: 2006 IEEE International Conference on Robotics and Biomimetics, pp. 601–606 (2006)
35. Tao, Y., Lam, T.L., Qian, H., Xu, Y.: A real-time intelligent shoe-keyboard for computer input. In: 2012 IEEE International Conference on Robotics and Biomimetics (ROBIO), pp. 1488–1493 (2012)
36. Ye, W., Xu, Y., Lee, K.K.: Shoe-Mouse: an integrated intelligent shoe. In: 2005 IEEE/RSJ International Conference on Intelligent Robots and Systems, pp. 1163–1167 (2005)

37. Paradiso, J.A., Hsiao, K., Benbasat, A.Y., Teegarden, Z.: Design and implementation of expressive footwear. IBM Syst. J. **39**, 511–529 (2000)
38. Papetti, S., Civolani, M., Fontana, F.: Rhythm'n'Shoes: a wearable foot tapping interface with audio-tactile feedback. In: NIME, pp. 473–476 (2011)
39. Matthies, D.J.C., Müller, F., Anthes, C., Kranzlmüller, D.: ShoeSoleSense: proof of concept for a wearable foot interface for virtual and real environments. In: Proceedings of the 19th ACM Symposium on Virtual Reality Software and Technology, pp. 93–96. Association for Computing Machinery, New York, NY, USA (2013)
40. Morley, R.E., Jr., Richter, E.J., Klaesner, J.W., Maluf, K.S., Mueller, M.J.: In-shoe multisensory data acquisition system. IEEE Trans. Biomed. Eng. **48**, 815–820 (2001)
41. Tao, Y., Qian, H., Yang, Y., Han, L., Xu, Y.: A real-time intelligent shoe system for writing by foot (2014). https://doi.org/10.1109/robio.2014.7090648
42. Tao, Y., Qian, H., Chen, M., Shi, X., Xu, Y.: A real-time intelligent shoe system for fall detection (2011). https://doi.org/10.1109/robio.2011.6181633
43. Sato, M.: Foot interface: fantastic phantom slipper. In: SIGGRAPH98–25th International Conference on Computer Graphics and Interactive Techniques, Orlando (1998)
44. Quek, F., Ehrich, R., Lockhart, T.: As go the feet … : on the estimation of attentional focus from stance. ACM Trans. Comput. Hum. Interact. **2008**, 97–104 (2008)
45. Paelke, V., Reimann, C., Stichling, D.: Foot-based mobile interaction with games. In: Proceedings of the 2004 ACM SIGCHI International Conference on Advances in Computer Entertainment Technology, pp. 321–324. Association for Computing Machinery, New York, NY, USA (2004)
46. https://play.google.com/store/apps/details?id=com.kkota323.socc. 29 Accessed April 2021
47. Paradiso, J., Abler, C., Hsiao, K.-Y., Reynolds, M.: The magic carpet: physical sensing for immersive environments. In: CHI '97 Extended Abstracts on Human Factors in Computing Systems, pp. 277–278. Association for Computing Machinery, New York, NY, USA (1997)
48. Fernström, M., Griffith, N.: LiteFoot—auditory display of footwork (1998). https://doi.org/10.14236/ewic/ad1998.13
49. Richardson, B., Leydon, K., Fernstrom, M., Paradiso, J.A.: Z-tiles: building blocks for modular, pressure-sensing floorspaces. In: CHI '04 Extended Abstracts on Human Factors in Computing Systems, pp. 1529–1532. Association for Computing Machinery, New York, NY, USA (2004)
50. Leikas, J., Väätänen, A., Räty, V.-P.: Virtual space computer games with a floor sensor control—human centred approach in the design process (2001). https://doi.org/10.1007/3-540-44589-7_22
51. Grønbæk, K., Iversen, O.S., Kortbek, K.J., Nielsen, K.R., Aagaard, L.: IGameFloor: a platform for co-located collaborative games. In: Proceedings of the International Conference on Advances in Computer Entertainment Technology, pp. 64–71. Association for Computing Machinery, New York, NY, USA (2007)
52. Chuang, L.-Y., Hung, H.-Y., Huang, C.-J., Chang, Y.-K., Hung, T.-M.: A 3-month intervention of dance dance revolution improves interference control in elderly females: a preliminary investigation. Exp. Brain Res. **233**, 1181–1188 (2015)
53. Hutton, E., Sundar, S.S.: Can video games enhance creativity? Effects of emotion generated by dance dance revolution. Creat. Res. J. **22**, 294–303 (2010)
54. Murphy, E.C.-S., Carson, L., Neal, W., Baylis, C., Donley, D., Yeater, R.: Effects of an exercise intervention using dance dance revolution on endothelial function and other risk factors in overweight children. Int. J. Pediatr. Obes. **4**, 205–214 (2009)
55. Trout, J., Zamora, K.: Using dance dance revolution in physical education. Teaching Elem. Phys. Educ. **16**, 22–25 (2005)
56. Sapiński, T., Kamińska, D., Pelikant, A., Anbarjafari, G.: Emotion recognition from skeletal movements. Entropy **21** (2019). https://doi.org/10.3390/e21070646

Chapter 6
Whole-Body Movement

6.1 Introduction

Whole-body movement can be used as a source of signals in human–computer interaction. Issues related to body recognition are of interest to researchers in fields as diverse as psychology, dance, and computer science (for instance, in affective computing).

6.2 Recognition of Body Movements

Recognition of users' movement in human–computer interaction involves identifying bodies (or their parts), and capturing and tracking their positions. Various technologies are used to capture and track motion, including motion detection video systems, motion trackers, and sensors fitted into mobile devices and touch screens. In human–computer interaction and, more broadly, in human-technology interaction, the emergence of methods that enable the recognition of users' movements has presented new opportunities to understand and enhance interactions, and created entirely new advanced interaction techniques [1, 2] and methods of manipulating objects in virtual 3D environments [3].

Body movement classification systems were first developed by Laban during his analyses of dance movement [4, 5]. The Labanotation, or Kinetography Laban system, derives from Laban Movement Analysis, which distinguishes four characteristics of movement: space, effort, shape, and body. Another important figure in the history of movement analysis was American anthropologist Ray Birdwhistell, the founder of kinesics: the study of body motion as a form of communication. This approach drew heavily on linguistics. Birdwhistell coined the term, "kineme": a unit that conveys meaning and is analogous to a phoneme in linguistics. The systems were based on a structural approach, concentrating on the identification of body

C. Biele, *Human Movements in Human-Computer Interaction (HCI)*,
Studies in Computational Intelligence 996,
https://doi.org/10.1007/978-3-030-90004-5_6

movements and the creation of a system that accurately describes them with the aim of allowing them to be reproduced. These systems served as the foundation of later computer-based movement description systems.

Presently, the most common approach to recognizing emotions from facial expressions—which is analogous to recognizing them from body movements—relies on disentangling complex movement patterns into smaller and more easily manageable elements. In the case of movement of the facial muscles, this is possible by virtue of the facial action coding system (FACS) [6], which has become the established starting point for solutions that recognize emotions based on facial expressions. Codes are defined and assigned to specific behaviors from a predetermined set, which enables avoidance of ambiguities and reduces the influence of observers' subjectivity, and increases the measurements' repeatability. Provided the conditions are ideal, assessments of the same material made by different judges should be identical. There are also systems tailored to specific research problems or communication channels—for example, gestures [7, 8]; they remain limited and susceptible to the subjective evaluations of the judges, however [9]. Systems based on the description of movement of individual limbs—such as the body action and posture coding system (BAP) of Deal et al. [10]—have proven more reliable. It is noteworthy, however, that, regardless of their form, systems in which coding is performed manually take inordinately long to configure. Moreover, the more detailed a system is (and the more actions it has to classify), the longer the coding process takes; for purposes of illustration, it takes up to 15 min to code a 2.5 s video from the GEMP database [11].

An effective body movement description system should be universal (able to code all forms of movement); comprehensive (include all aspects of movement); allow for the analysis of movement; describe movement with regard to its intention; be adaptable to various applications; be logical and practical; and facilitate easy integration with technological solutions [12]. It should also ensure the objectivity of assessments without employing assumptions on the movement assessed.

6.2.1 Body Movements and Affective Computing

Although body movements convey less information on emotional states than facial expressions do [13], they can nevertheless be used as a source of information in human–computer interaction. Owing to their inextricable link, the analysis of body movements offers valuable data on users' emotional states; in consequence, serves the design of affective computing systems, which focus on emotions [14] and allow effective monitoring of emotional states to improve the quality of communication between humans and machines. Facial expressions and psychophysiological signals are the most widely studied issues in this context (see other chapters of this volume). Relatively little attention has been paid, however, to the emotional signals that manifest in body movements.

Data-driven solutions may rely on data from motion capture systems or on those that gather information on the positions of individual body parts. Data collected in this way must then be dissected into individual recorded activities, which can be performed with the support of computer systems—for example, using the graph cut method [15]. The data is then used to train classifiers, which can employ logistic regression or decision trees [16]. The knowledge-driven approach—which relies on the theories and knowledge of psychology, anthropology, or the arts (specifically, dance)—is an alternative. Such systems typically comprise solutions based on Labanotation [17] or BAP [10]. Other systems are specifically designed to recognize emotions, in which movement classification is seen merely as a means of achieving this goal [18].

6.2.2 Automatic Recognition of Body Movements

The number of technological solutions designed to automatically recognize emotional states on the basis of movement of the body (or its parts) is significantly lower than the number designed to recognize emotions from facial expressions. Current solutions typically rely on video-based recognition or on data from motion capture systems.

6.2.3 Automatic Recognition Using Motion Capture Systems

Initial attempts to use data from motion capture systems to recognize emotions on the basis of body movements were made by Kapur et al. [16]. They utilized a motion capture system with 14 reference points. Subjects were asked to express four emotions (sadness, happiness, anger, and fear). In order to obtain ground truth, the recorded videos were classified by judges. The feature extraction process used the positions, velocities, and accelerations that were able to be determined for each of the reference points; for each of these indicators, the means and standard deviations were then calculated over ten seconds for each emotion depicted. Five classifiers were used in the machine learning experiments: a logistic regression, a naive Bayes, a decision tree, a multi-layer perceptron backpropogation artificial neural network, and a support vector machine. Their effectiveness ranged between 66% (naive Bayes) and 92% (the support vector machine)—human judges achieved a recognition rate of 93% when shown the recorded clips. It is worth remembering, however, that the stimuli were videos in which the actors intentionally demonstrated specific emotions. The question remains whether it is possible to read emotions from unintentional behaviors—a problem that is far more complex than it appears. The analysis of spontaneous behavior must account for several factors that are negligible when analyzing intentional demonstrations of emotion. It is necessary, for example, to identify which elements of movement are related to emotions, and to distinguish them from those that result from individual factors related to the actors—such as personality traits or gender. This issue is well known in computer animation: the same action performed by characters (for instance, opening a door) must look different depending on their emotional states; two characters will never perform an identical action [19]. Systems that generate virtual characters treat emotions as an additional layer that changes the manner in which the characters behave—for example, a movement's amplitude or speed [20, 21]. In the case of automatic recognition of emotions based on movement, this layer must be detached from the movement and processed further to classify the emotional state.

Bernhardt and Robinson attempted to establish whether it was possible to automatically read emotions from the natural movements of bodies [22]. They used a publicly available motion capture database [23] containing recordings of various types of movement, such as knocking or picking up objects, performed in different emotional states (neutral, happy, angry, and sad). Movement data was recorded for 15 body points. The analyses were based on data from the recordings of knocking movements. First, the data was processed to eliminate the potential impact of scale and rotation—the reference point was a coordinate system based on the users' bodies (and not on the room). The data was then normalized with regard to the sizes of the bodies. Next, complex multi-stage movements were reduced into simpler components. This step was based on determining the movement energy level for each data point by determining the angular velocity of each sensor. Then, based on an agreed cut-off point—the energy value above which the movement could be classified as occurring at a given moment, and below which it could not—was established. Each point in time was then classified as either movement or stationary points. As a result of this classification, the fragments of data that formed the individual components of complex movement patterns were identified. It is possible to employ data clustering methods to combine the components into larger groups connected by meaning—for example, phases. For the knocking movement, Bernhardt and Robinson determined four so-called "motion primitives"—raising a hand, lowering a hand, withdrawing a hand, and knocking. The process of recognizing emotions on the basis of movement is possible only following thorough analysis and appropriate classification.

The two movement parameters that are the most closely related to emotions are speed and acceleration [16, 24, 25]. In the case of hand or arm movements, such as knocking, mean velocity, mean acceleration, mean jerk, and distance from the body can serve as effective indicators. In order to minimize the influence of individual differences on movement (each person performs a given movement uniquely, depending on their psychological and physical traits), it is necessary to standardize the collected data—or, in other words, to accept the average value of all accessible recordings for a given individual as a reference point. This procedure is analogous to the standard one used on psychophysiological data, such as that of muscle activity, recorded with the use of surface electromyography (EMG). [26]. In this case and in that of movement data analysis, the application of intra-person standardization minimizes the impact of individual factors, in addition to enabling researchers to compare individuals' data directly. The standardized data can then be subjected to machine learning procedures—with the use of support vector machines, for instance.

Bernhardt and Robinson's results demonstrate that the effectiveness of classification, without exclusion of individual factors, is considerably lower than it is following standardization—in which the recognition rate is higher than that achieved by humans in the tasks that involve recognizing movement on the basis of point-light displays [27].

6.2.4 Automatic Recognition Using Video Data

In the case of systems that use video signals, the recognition process is more complex, as movement classification must be preceded by video signal processing. Camurri et al. propose dividing the process of recognizing emotions on the basis of body movements into four layers.

In the first layer, video material is processed frame by frame to extract information on the movements being performed. Computer vision algorithms based on detection of changes over time [28] and trait tracking [29] can be used for this purpose. The outputs of the signal processing in the first layer are processed images and the movement trajectories of individual body parts.

In the second layer, motion cues describing the movement and its parameters are determined based on the data containing the images and trajectories from the first. Two indicators are used for this purpose: quantity of motion (QoM) and contraction index (CI). The QoM indicator allows the amount of motion to be approximated. It utilizes Silhouette Motion Images (SMIs), which integrate the movement of the silhouettes visible across several video frames. The comparison between the area occupied by the silhouettes at a given moment and the SMIs from several previous frames offers an accurate indication of momentum—the product of body mass and body velocity. The CI, on the other hand, is comparable to Laban's "personal space". To calculate the CI, the actual and minimum silhouette sizes (the smallest possible rectangle in which a silhouette is inscribed) must be compared. This procedure allows researchers to determine whether a silhouette comprises one or several solids. Comparing the state at the beginning and at the end of movement enables detection of the direction of postural changes. The processing of data in the second layer involves a number of methods used in computer vision, signal processing, and statistics (Fig. 6.1).

Automatic recognition systems must also capture every stage of movement and detect motionlessness. These processes are performed in the third layer. Moments in which an individual remains motionless can be detected by analysis of the QoM—readings below a predetermined threshold indicate motionlessness. In the case of longer movement sequences, capturing moments of movement and motionlessness allows researchers to determine the flow of movement. The analysis of QoM changes over time facilitates the detection of fast movements, in which velocity increases rapidly from zero to the maximum; and for slow ones, in which velocity also changes slowly.

In the fourth layer, which integrates information from the second and third layers, movement is classified by basic emotions. This classification can be performed using a variety of methods drawing both from statistics (for instance, regression)

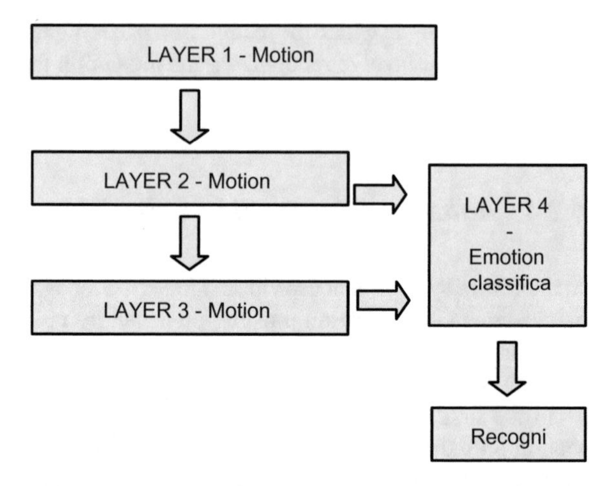

Fig. 6.1 The layered approach described by Camurri et al. [24]

and machine learning (for instance, support vector machines or neural networks). It is worth remembering that the division into movement analysis layers outlined above was proposed with the purpose of studying the emotions expressed during dance. Despite this specificity, its assumptions appear universal, and could potentially be applied in analyses on the extraction of emotions from other forms of movement.

6.3 Practical Application of Body Movement Recognition

6.3.1 Interpretation of Sign Language

Motion recognition with the use of Microsoft's X-Box Kinect system has been tested as a potential method of enhancing interactions between sign language users—similarly to how voice recognition is used in the case of those without hearing and/or speech impairments. Oszust and Wysocki [30] tested two types of data provided by the Kinect system—skeletal images of the body and images of hand positions extracted from a video signal as skin colored regions. Polish Sign Language (PJM) was used to conduct the tests. In sign language, signs are frequently expressed in a highly dynamic manner, and the hands often cover and touch each other. This results in a number of problems, even during the first stage of the recognition process, hand segmentation. This serves as the basis of the feature extraction process, in which information on various features is collected, such as the location of the gravity center of each hand in relation to the position of the face, and the number of pixels covered by the hands. The data obtained must then be combined with depth data provided by the Kinect sensors. This is because when communicating in sign language, the hands are regularly placed in front of the neck, which is of the same color; additionally, they are regularly placed one in front of the other. The tests with video recordings of 30 words have demonstrated that methods that combine hand images with depth data yield the highest recognition rates.

6.3.2 *Person Identification*

The analysis of body movements and the shape of the body can also prove useful in identification. Systems that are capable of performing such operations are being developed because, unlike other popular methods of biometric identification, such as fingerprints, they can be used without interaction with nor the knowledge of the individual being identified. Anthropometric analyses with the purpose of identifying individuals unequivocally were conducted before computer methods were available, including in forensics. More recently, the possibility of using computer methods to perform such tasks was studied by Godil et al. [31]. Using the publicly available CEASAR database [32], which contains anthropometric data on several thousand individuals in Europe and North America, the researchers studied identification based on analysis of the sizes of body parts extracted from photographs.

Humans differ not only in the size of their bodies, but also in their gaits. Gait analysis could prove an effective method of identification, insusceptible to the typical interference resulting from, for example, changes in mood or the wearing of accessories, such as backpacks. Munsell et al. [33] used Kinect depth sensor recordings to analyze the relative movements of the upper and lower halves of the body; their identification rate was 90%.

6.4 Body Movement as Input for Interactions

Body movement can be used not only as a source of signals in ion detection algorythms i.e. affective computing, but also as a means of control in human–computer or human–robot interaction.

6.4.1 *Remote Operation of Non-humanoid Robots*

Remote operation is a field of science that focuses on the remote control of devices. Its history dates back to 1877, when remote-controlled torpedoes were patented by Luis Brennan. Several years later, Nikola Tesla developed the first boat that was controlled remotely using radio waves.

Current solutions that allow the remote control of robots rely primarily on systems based on arbitrary assignment of the actions of operators to those of the robots they control. Operators' actions are determined by the characteristics of the control interface, which, in many cases, leads to control over the robot becoming, in itself, a challenging skill to master. One solution involves utilizing natural body movements; another involves brain-computer interfaces that bypass the behavioral path, and that are based on data read directly from the activity of the cerebral cortex.

6.4.2 Body-Machine Interfaces

Body-machine interface systems can be useful when the use of standard computer interfaces proves impracticable, such as when partially paralyzed users are involved [34, 35]. These systems collect signals from body movements, which is significantly easier than doing so directly from the cerebral cortex (as is the case with brain-computer interfaces). The most appreciable advantage of body-machine interfaces over brain-computer interfaces is that they can use motor ability (even when partially impaired) to control new actions in a controlled interface. Brain-computer interfaces demand conscious control of brain activity—a skill that is arduous to acquire (see Future Directions chapter).

The body-machine interface approach can be effective for those who, despite their injuries, remain capable of performing certain motor actions. Its effectiveness has been demonstrated by an artificial arm controlled using EMG signals [36] and a system that enables its users to control wheelchairs using their tongues [37]. Abdollahi et al. [34] devised a system to improve the quality of life of those who have suffered spinal injuries, but have not fully lost their motor skills. It relies on inertial sensors attached to a specially-constructed vest (four sensors ensure motion capture in eight degrees of freedom); the resulting signals are converted into a two-dimensional task space. The effectiveness of the system was tested in different tasks: cursor control, typing on a virtual keyboard, and solitaire games. Tests conducted by the authors indicate that body-machine interface solutions can serve as effective computer interaction tools for those who have partially retained mobility following spinal injuries.

6.4.3 Control of Drones Using Body-Machine Interfaces

Use of the body as a source for control signals reaches beyond the medical sphere; whole-body movement can also be used to control drones. These flying robots have a host of useful applications, which underlines the importance increasing their efficiency, ease of control, and the length of operator training required for them to be flown safely.

Solutions outlined in the literature rely both on signals from the whole body and from its parts. Pittman and LaViola [38] compared different methods of controlling drones using head movements, such as shifting and rotating, as well as more conventional methods, including the use of Nintendo Wiimotes. Their effectiveness was tested by measuring the time required to navigate a drone along a set route using each method. Subjectively assessed interactions were also examined. The subjects completed their tasks the fastest using the Wiimote controller. Rotating the head proved more effective than shifting it. In ease of use, the traditional controller method was again rated the highest. The experiment shows that the task of designing effective systems that translate head movements into those of remote-controlled drones is a laborious one. The low efficiency of Pittman and LaViola's system might be explained by the counterintuitive head movements that were selected for steering. Based on a similar assumption, Miehlbradt et al. [39] conducted preliminary studies to determine what movement strategies (and body movements) subjects spontaneously used to control drones. By measuring the activity of various muscle groups in the upper body—specifically, the arms and torso—the authors observed that the subjects used two strategies based on separate movement patterns: one involved only torso movements, while the other involved both the torso and the arms. This strategy enabled the authors to devise a control method based on users' most common movement patterns, which was highly effective in training—it allowed the subjects to learn how to control the drone faster than traditional methods.

6.4.4 *Virtual Mouse*

The Kinect system is a peripheral device for Microsoft's Xbox 360 console, comprising an RGB camera, an infrared image camera, an infrared light transmitter, and an audio recorder. This configuration allows the system not only to obtain color video images, but also depth maps, which detect the distance between the camera and the objects being filmed. Kinect also processes information on the 3D positions of 20 major joints. Technical parameters such as these coupled with relatively simple operation, from a programming perspective, have contributed to the system's popularity beyond Xbox games; Kinect also serves as a base for the creation and testing of innovative solutions in human–computer interaction via whole-body movement [30, 40–42].

Additionally, the Kinect sensor was used as an interaction tool in the design of a virtual mouse by Szeghalmy et al. [40]. For practical reasons, the role of Kinect was limited to the processing of hand gestures, rather than full-body movement. The system recognizes five gestures that were assigned, either individually or in combinations, to a series of actions—including start, stop, click, and double click. Preliminary tests on the system's effectiveness have yielded promising results. The authors admit, however, that users operating the device for the first time often find it cumbersome and counterintuitive. Moreover, the small sample size fails to elucidate clear conclusions. Similar experiments were conducted by Osunkoya and Chen [42]. They adapted the Kinect system to allow control of computer presentations. Appropriate hand gestures are interpreted by the system as keystrokes that allow the user to toggle slides.

6.5 New Paradigms of Body-Movement-Based Interaction

Aside from attempts to apply the paradigms of traditional solutions in human–computer interaction—such as the virtual gesture-controlled mouse described above—entirely innovative methods of interaction are being established. As solutions like the Nintendo Wii controller or the Kinect system use body movement as sources of interaction data, it is now possible to devise and test new approaches to interaction with computers—for example, those involving the whole body. Bianchi-Berthouze et al. [43] claim—based on their research on the relationship between body movement and involvement in interaction—that movement of the whole body enhances feelings of immersion in virtual environments, facilitates emotional communication, and improves the functioning of emotional regulatory mechanisms.

6.5.1 *Body Menu*

One example of such a system is Body Menu by Bossavit et al. [2]. Individual actions of the system are paired with separate areas of the body. To activate a function, a user's hand must be placed in a designated area of the body—for example, on the left side of the chest, or at the height of the belly button. Its effectiveness was tested against two other solutions in which the active areas were located outside the users' bodies. User studies demonstrated that Body Menu was the easiest to use.

6.6 Conclusions

Systems that use whole-body movement signals were, until recently, highly complex and required specialized equipment. With the development of immersive VR, the prospects for such interactions have become significantly more promising. The most rudimentary VR systems typically allow users to track three key points in space—the head and both of the hands (controllers). They are also easy to expand using additional sensors, which enables the systems to obtain accurate data on the positions of their users' bodies. For these reasons, it is widely expected that new body-movement-based interaction methods designed for immersive VR will be developed in the future.

References

1. Perelman, G., Serrano, M., Raynal, M., Picard, C., Derras, M., Dubois, E.: The roly-poly mouse: Designing a rolling input device unifying 2D and 3D interaction. In: Proceedings of the 33rd Annual ACM Conference on Human Factors in Computing Systems, pp. 327–336. Association for Computing Machinery, New York, NY, USA (2015)

2. Bossavit, B., Marzo, A., Ardaiz, O., Pina, A.: Hierarchical menu selection with a body-centered remote interface. Interact. Comput. **26**, 389–402 (2013)
3. Bossavit, B., Marzo, A., Ardaiz, O., De Cerio, L.D., Pina, A.: design choices and their implications for 3D mid-air manipulation techniques (2014). https://doi.org/10.1162/pres_a_00207
4. Laban, R.: Laban's principles of dance and movement notation principles of dance and movement notation. Plays, Boston (1975)
5. von Laban, R.: Principles of dance and movement notation: with 114 basic movement graphs and their explanation. Macdonald & Evans (1956).
6. Ekman, R.: What the face reveals: basic and applied studies of spontaneous expression using the facial action coding system (FACS). Oxford University Press (1997)
7. Lausberg, H., Sloetjes, H.: Coding gestural behavior with the NEUROGES–ELAN system. Behav. Res. Methods. **41**, 841–849 (2009)
8. Cohen, D., Beattie, G., Shovelton, H.: Nonverbal indicators of deception: how iconic gestures reveal thoughts that cannot be suppressed. Semiotica. (2010). https://doi.org/10.1515/semi.2010.055
9. Scherer, K.R., Ekman, P.: Methodological issues in studying nonverbal behavior. Handbook of Methods in Nonverbal Behavior Research. 1–44 (1982).
10. Dael, N., Mortillaro, M., Scherer, K.R.: The body action and posture coding system (BAP): development and reliability. J. Nonverbal Behav. **36**, 97–121 (2012)
11. Bänziger, T., Scherer, K.R.: Introducing the geneva multimodal emotion portrayal (gemep) corpus. Blueprint for Affective Computing: A Sourcebook **2010**, 271–294 (2010)
12. Guest, A.H.: Dance notation: the process of recording movement on paper. Dance Horizons, New York (1984)
13. Ekman, P.: Differential communication of affect by head and body cues. J. Pers. Soc. Psychol. **2**, 726–735 (1965)
14. Picard, R.W.: Affective computing (2000)
15. Yu, X., Liu, W., Xing, W.: Behavioral segmentation for human motion capture data based on graph cut method (2017). https://doi.org/10.1016/j.jvlc.2017.09.001
16. Kapur, A., Kapur, A., Virji-Babul, N., Tzanetakis, G., Driessen, P.F.: Gesture-based affective computing on motion capture data. In: Affective Computing and Intelligent Interaction, pp. 1–7. Springer Berlin Heidelberg (2005)
17. Samadani, A., Burton, S., Gorbet, R., Kulic, D.: Laban effort and shape analysis of affective hand and arm movements. In: 2013 Humaine Association Conference on Affective Computing and Intelligent Interaction, pp. 343–348 (2013)
18. Coan, J.A., Gottman, J.M.: The specific affect coding system (SPAFF). Handbook of Emotion Elicitation and Assessment, p. 267 (2007)
19. Turner, P.: The language of cinema and traditional animation in the 3D computer animation classroom (1998). https://doi.org/10.1145/280953.281010
20. Sannier, G., Balcisoy, S., Magnenat-Thalmann, N., Thalmann, D.: VHD: a system for directing real-time virtual actors (1999). https://doi.org/10.1007/s003710050181
21. Zhao, L., Costa, M., Badler, N.L.: Interpreting movement manner. In: Proceedings Computer Animation 2000, pp. 98–103 (2000)
22. Bernhardt, D., Robinson, P.: Detecting affect from non-stylised body motions. In: Affective Computing and Intelligent Interaction, pp. 59–70. Springer Berlin Heidelberg (2007)
23. Ma, Y., Paterson, H.M., Pollick, F.E.: A motion capture library for the study of identity, gender, and emotion perception from biological motion (2006). https://doi.org/10.3758/bf03192758
24. Camurri, A., Lagerlöf, I., Volpe, G.: Recognizing emotion from dance movement: comparison of spectator recognition and automated techniques. Int. J. Hum. Comput. Stud. **59**, 213–225 (2003)
25. Pollick, F.E., Lestou, V., Ryu, J., Cho, S.-B.: Estimating the efficiency of recognizing gender and affect from biological motion. Vision Res. **42**, 2345–2355 (2002)
26. Stålberg, E., van Dijk, H., Falck, B., Kimura, J., Neuwirth, C., Pitt, M., Podnar, S., Rubin, D.I., Rutkove, S., Sanders, D.B., Sonoo, M., Tankisi, H., Zwarts, M.: Standards for quantification of EMG and neurography. Clin. Neurophysiol. **130**, 1688–1729 (2019)

27. Pollick, F.E., Paterson, H.M., Bruderlin, A., Sanford, A.J.: Perceiving affect from arm movement. Cognition **82**, B51-61 (2001)
28. Bobick, A.F., Davis, J.W.: The recognition of human movement using temporal templates (2001). https://doi.org/10.1109/34.910878
29. Lucas, B.D., Kanade, T., Others: An iterative image registration technique with an application to stereo vision (1981)
30. Oszust, M., Wysocki, M.: Recognition of signed expressions observed by Kinect sensor. In: 2013 10th IEEE International Conference on Advanced Video and Signal Based Surveillance, pp. 220–225 (2013)
31. Godil, A., Grother, P., Ressler, S.: Human identification from body shape. In: Fourth International Conference on 3-D Digital Imaging and Modeling, 2003. 3DIM 2003. Proceedings, pp. 386–392 (2003)
32. Robinette, K.M., Blackwell, S., Daanen, H., Boehmer, M., Fleming, S.: Civilian American and European Surface Anthropometry Resource (CAESAR), final report, vol. 1. Summary. Sytronics Inc Dayton Oh (2002)
33. Munsell, B.C., Temlyakov, A., Qu, C., Wang, S.: Person identification using full-body motion and anthropometric biometrics from Kinect videos.In: Computer Vision—ECCV 2012 Workshops and Demonstrations, pp. 91–100. Springer Berlin Heidelberg (2012)
34. Abdollahi, F., Farshchiansadegh, A., Pierella, C., Seáñez-González, I., Thorp, E., Lee, M.-H., Ranganathan, R., Pedersen, J., Chen, D., Roth, E., Casadio, M., Mussa-Ivaldi, F.: Body-machine interface enables people with cervical spinal cord injury to control devices with available body movements: proof of concept. Neurorehabil. Neural Repair. **31**, 487–493 (2017)
35. Casadio, M., Pressman, A., Acosta, S., Danzinger, Z., Fishbach, A., Mussa-Ivaldi, F.A., Muir, K., Tseng, H., Chen, D.: Body machine interface: remapping motor skills after spinal cord injury. IEEE Int. Conf. Rehabil. Robot. **2011**, 5975384 (2011)
36. Kuiken, T.A., Li, G., Lock, B.A., Lipschutz, R.D., Miller, L.A., Stubblefield, K.A., Englehart, K.B.: Targeted muscle reinnervation for real-time myoelectric control of multifunction artificial arms. JAMA **301**, 619–628 (2009)
37. Huo, X., Ghovanloo, M.: Using unconstrained tongue motion as an alternative control mechanism for wheeled mobility. IEEE Trans. Biomed. Eng. **56**, 1719–1726 (2009)
38. Pittman, C., LaViola, J.J., Jr: Exploring head tracked head mounted displays for first person robot teleoperation. In: Proceedings of the 19th International Conference on Intelligent User Interfaces (2014)
39. Miehlbradt, J., Cherpillod, A., Mintchev, S.: Data-driven body–machine interface for the accurate control of drones. Proceedings of the National Academy of Sciences (2018)
40. Szeghalmy, S., Zichar, M., Fazekas, A.: Gesture-based computer mouse using Kinect sensor (2014). https://doi.org/10.1109/coginfocom.2014.7020491
41. Saha, S., Ganguly, B., Konar, A.: Gesture based improved human-computer interaction using microsoft's Kinect sensor (2016). https://doi.org/10.1109/microcom.2016.7522582
42. Osunkoya, T., Chern, J.-C.: Gesture-based human-computer-interaction using Kinect for windows mouse control and powerpoint presentation. In: Proceedings of 46th Midwest Instruction and Computing Symposium (MICS2013), pp. 19–20 (2013)
43. Bianchi-Berthouze, N.: Understanding the role of body movement in player engagement. Hum. Comput. Interaction. **28**, 40–75 (2013)

Chapter 7
Movement in Virtual Reality

7.1 Introduction

We are on the verge of a revolution during which human functioning in societies and communities will be changed irreversibly; this will be driven by immersive virtual reality (IVR), and will have a similar, or even greater, impact than that of the internet or smartphones. The first publicly available devices that allow users to experience the synthetic world have been launched in recent years. At the end of 2019, the plans of one technology giant to create a social network that operates in VR were released and in 2021 they were confirmed. Another giant - Apple plans to enter the market in 2022. From this perspective, studies on how humans will function socially in the new virtual world are of particular relevance. Similar questions about the "real" world have been answered by the psychology of emotions and of gender differences. Knowledge of these new areas, however, remains limited. A unique property of IVR is its allowance of relative freedom of movement, which, in turn, allows the movement of individual parts of users' bodies to be analyzed. To facilitate effective analysis of such large amounts of data, it will likely be necessary that methods traditionally used in the big data domain be applied. It can be expected that the use of machine learning methods, in conjunction with the experimental psychology of emotions, will facilitate the creation of a unique classifier of body movement patterns that are characteristic for individual emotions experienced in IVR. Such a tool will enable researchers and practitioners to easily collect and analyze data on the emotions of their subjects, which will help them to better understand the psychological functioning of individuals in IVR; thus, paving the way for modern research methods in psychology, diagnostics, and the therapy of emotional disorders.

C. Biele, *Human Movements in Human-Computer Interaction (HCI)*,
Studies in Computational Intelligence 996,
https://doi.org/10.1007/978-3-030-90004-5_7

7.2 Psychological Research in Virtual Reality

Given the plethora of scientific papers published on personality psychology and movement analysis, surprisingly little of the research pertains to individual differences in movement. The idea that personality traits are observable through behavior was first observed in antiquity by Greek philosophers, such as Theophrastus, who defined types of people by the behaviors they exhibited. Sheldon refined the definition of behavior in his contemporary classical works on the psychology of temperament. More recent works by Eysenck, the architect of the most widespread theory of personality and personality questionnaires [1], rest on the assumption that the root cause of differences in behavior—for instance, in the manner of movement—of different personality types (which is most evident in the opposition extraversion) is biological. Although behaviors, reactions, and mobility are viewed as important factors that relate to individual traits, they have not been subjected to in-depth research. This might be explained by the technical difficulty involved in conducting objective studies of subjects' movements. The little information on the matter that can be found in the literature concerns only specific types of movement. The reports that can be found are based on highly invasive methods, such as the study of eyeball movement or mobility measured using smartphone accelerometers; utilize subjective measures [2]; or study behaviors in highly artificial laboratory environments, such as moving a ball on a string [3]. Despite their limitations, the studies suggest that it is possible to infer personality traits from movement analyses and related data. A need remains for a method that could non-invasively and subjectively collect motor data from subjects, facilitating further analyses.

The term, virtual reality is often used interchangeably with three types of computer system: flat-screen virtual environments; room and wall-mounted systems, such as CAVE; and those that utilize head-mounted displays (HMDs) and controllers. The last is of an entirely different quality. HMD systems evoke the strongest impression of immersion, allow for free movement in virtual worlds, and offer the tracking and mapping of the positions of the head and hands—and, in some solutions, the reading of the position of the eyes and the objects of their focus. This changes the way we experience virtual environments profoundly [4]. They are known as immersive virtual reality systems to distinguish them from general virtual reality. Ease of use and low costs make consumer HMD systems, such as the Oculus Rift S or the HTC Vive models, the only solutions with the potential for mass use in psychological research and therapy. Most importantly from the perspective of this chapter, consumer HMD systems record the positions of users in both virtual and physical worlds, and the position and orientation of specific body parts (including the head, hands, and eyes) with high spatial accuracy and time resolution. Resultantly, they are able to collect objective data without the need for additional equipment.

7.2.1 *Emotions are Observable in Movement*

The relationship between emotion and movement has been explored in research much more frequently than that between movement and personality. This started with Ekman's classic studies on the mimical expression of emotions; continued with studies on the expression of emotions using posture, on the role of movement and gender differences in the recognition of emotions [5], and on the neurophysiological elements of emotional mimicry; and led to the latest research that suggests the existence of direct connections between the brain's limbic and motor structures [6]. Attempts have also been made to infer emotions on the basis of motion analysis—by analyzing images from cameras, for instance. No study to date, however, has analyzed movement and motion data in IVR in relation to emotions. It is noteworthy that despite exhaustive exploration of gender differences in the processing of emotions in psychology, there is currently no data that pertains to them in IVR. Works summarizing the results of over a century of research into how significant and fundamental the differences are were published in the 1970s and 1980s ([7, 8]). Contrary to popular assumptions of sex differences, in which women are more emotionally sensitive than men, empirical research paints a more complex picture. Women react to emotional stimuli more expressively than men, confirm that they feel more emotional, and display elevated physiological responses [9]. Reports on this subject remain inconsistent, however, and the magnitude of sex differences observed varies considerably between studies. Current meta-analyses [10] even suggest that they might be less considerable than previously thought.

7.2.2 *Machine Learning Can Be Used as a Tool to Analyze Human Motion Data*

The data that characterizes human motion is multidimensional, heterogeneous, and non-linear; it often fails to meet the requirements of parametric tests, meaning that it might prove impracticable to use standard methods of statistical analysis in this context. Advanced analytical techniques from research on artificial intelligence work much better when applied to such data. Using them for user motion data in IVR could enable a breakthrough in psychological research, similar to the ones experienced in robotics and autonomous car research, speech and image recognition, and automatic cancer detection. Current studies, mainly in biomechanics and rehabilitation [11, 12], indicate that machine learning is an effective method for analyzing movement data; the main obstacle to its widespread use is the difficulty in collecting large amounts of reliable data. In the case of motion, this usually involves using a large number of external sensors or video recording—both of which entail serious limitations and are prone to data gaps. When machine learning is used to analyze motion data in IVR to recognize states and psychological characteristics, such obstacles disappear.

7.2.3 Differences Between Men's and Women's Behavior in IVR

Emotions, moods, and sentiment have long been established as factors that shape the course of users' interactions with computers [13, 14]. More recent research indicates that emotional stimuli presented during VR interactions might potentiate emotions differently from those presented on computer screens [15]. Research into the role of sex differences (particularly emotional ones) that occur during computer interactions continues to be extremely rare. This applies particularly to its application in the use of VR. The few reports available indicate that sex can influence presence [16] and how users navigate virtual worlds [17]. The quoted research serves as a key source of knowledge on sex differences in the reception and use of VR. It should be remembered, however, that these studies were not conducted in IVR, which differs considerably from 3D environments (these are also referred to as VR in older studies) presented on computer screens. The use of advanced technology that allows free (limited only by the size of rooms) movement in virtual worlds and the presentation of those worlds in three-dimensional form via HMDs deeply alters users' experiences of virtual worlds [4].

7.2.4 Personality Traits and Behavior in Immersive Virtual Reality

Studies on the relationship between personality traits and behavior in IVR are scarce. A review paper published in 2018 [18], which analyzed research on the relationship between personality traits and behavior in virtual reality, cited only twenty scientific papers. These were selected among almost 400 others that had been retrieved initially from the IEEE Xplore, ACM Digital Library and PubMed databases on the basis of fulfillment of the thematic criterion. Such a meagre number of relevant papers indicates clearly that the subject is worthy of further research and analysis. Moreover, almost none the studies found and analyzed by Aranhe et al. had been implemented in IVR; of the twenty, only one used an HMD: Oculus Rift. Interestingly from the perspective of this chapter, the authors of the systematic review failed to uncover any paper that suggested solutions to the problems of automatic recognition and prediction of personality—despite the existence of literature on the subject. Most of the studies conducted to date—whether listed in the review or the few that reach beyond it [19, 20]—abide by the correlational or differential paradigm, comparing the questionnaire measurement of personality traits with behavior in VR; effectively, they identify relationships between one and the other. Another paradigm focuses on examining whether individuals are able to recognize personality traits based on the head and eye movement of humanoid avatars. The available literature [21–23] confirms that they are.

7.2.5 *Measuring Brain Activity During Movement in IVR*

In psychological research, data is collected using three main types of measurement. The first, and by far the most widespread, is based on data that is sourced directly from the respondents—in the form of questionnaires or interviews, for example; the second is based on behavioral data derived from observation of behaviors in the laboratory or in the field; the third measures physiological parameters to record data on the body's reactions, such as changes in heart rate or the degree of muscle-tension associated with mental processes. Each of these methods has a number of disadvantages: self-description hinges on individuals' awareness (or lack thereof) of their own traits or behaviors, and is distorted by some respondent's desire to present themselves favorably; observation demands rigorous training of the observer to ensure its reliability, and requires the creation of controllable test procedures; physiological measurements are intrusive, and require electrodes to be connected to the body, which affects the behavior of subjects and decreases the accuracy of test results. All of the methods remain imperfect, despite their century-long development.

Functional near-infrared spectroscopy (fNIRS) is a novel brain activity imaging technique that allows relatively non-invasive and portable measurement of oxygenated blood flow. It utilizes radiation with a wavelength close to the infrared range. Light sources and detectors can recognize when radiation penetrates scalp tissues; the intensity depends on the amount of hemoglobin that is oxygenated, which, in turn, indicates the activity of neurons for which oxygen is supplied to the blood. Compared to other brain activity imaging methods, fNIRS is less restrictive, more convenient, and more portable. This advantage is of particular relevance in studies of emotional reactions conducted in IVR. fNIRS is currently the only method of recording brain activity that is insusceptible to motion artifact, as it entails no restrictions on the movement of subjects using portable equipment. It is known from the literature that fNIRS can be effective in recognizing and differentiating emotions [24], and in determining their valence [25]. It is also used in conjunction with immersive virtual environments to detect cognitive load, for instance [26].

7.2.6 *Measuring Visual Attention and Spatial Behavior in Virtual Reality*

Each user who participates in a VR experience can decide how he/she will move and on what his/her visual attention will focus. In experiences in which it is useful to control users' attention or behavior, it may also prove necessary to record data pertaining to their attention and movement paths—for example, to assess the effectiveness of indicators. Eye-tracking time series data of oculomotor activity can be used as visual attention trails; positional time series data can be used as spatial behavior trails. The distance correlation coefficient [27] can be used to measure the correspondence of users' experiences in VR [28]—both between different users and between experiences of the same user (to measure the consistency of their behavior over time, for instance). Data can be obtained from headsets and additional sensors. To effectively measure interpersonal dependence or intra-personal independence, a correlation measure that is appropriate for the data collected must be used. Measures should allow for the detection of as many correlations as possible, and the possibility to state that there is no correlation. From the perspective of data streams recorded during VR experiences—which are multidimensional and are collected from various types of sensor—a correlation measure appropriate for use in VR must meet several conditions: first, it should be capable of analyzing multi-dimensional data; second, it should be sensitive to non-linear relationships between variables; third, it should assume independence of individual data streams; fourth, it should allow null hypotheses to be tested; and fifth, it should be usable with time series data. All of these conditions are met by distance correlation. Its use has also been recommended by Kobyliński et al. [28].

7.2.7 *Virtual Reality as a Tool for the Treatment of Disorders*

VR systems are applied not only in psychological studies, but also in psychological and psychiatric practice. Since VR has the ability to present fairly accurate images of the real world and precise measurements of behavior, it can prove useful in the study of phobias, such as social anxiety or arachnophobia, and in addiction recovery programs.

According to the criteria adopted by the American Psychiatric Association in the Diagnostic and Statistical Manual of Mental Disorders (DSM), the constituents of phobias include fear, physiological reactions, and behavioral avoidance reactions caused by stimuli or situations that cause phobias. This type of disorder is diagnosed using the Behavioral Avoidance Test (BAT) [29]. Test subjects enter a room with a table located approximately 3 m in front of them. A closed jar containing a medium-sized living spider is placed on the table. The subjects' task is to approach the spider and, if possible, to take it in their hand after opening the jar. The experimenter does not encourage the subjects to undertake any of the steps, and the results are expressed on a 10-point scale, ranging from 1 (reaching a distance of 2 m from the spider) to 10 (taking the spider in the hand). This procedure was recreated in VR by Muhlberger et al. [30]. The researchers conducted tests on individuals who suffered from arachnophobia and measured their behavioral reactions, such as the distance maintained between subjects and the virtual spider, and electrodermal (skin conductance) levels. Their results clearly indicate that VR environments that contain anxiety stimuli can evoke verbal, physiological, and behavioral reactions characteristic of those who suffer from arachnophobia.

VR has also been postulated as a tool that can be useful in the case of alcohol use disorder (AUD). Dual-process models of addictive behavior [31] assume the existence of two mental processes that compete and determine the reaction of humans to stimuli in their environments. The first is the reflective process, which determines conscious behavior; the second is the impulsive process, which controls behavior automatically. The relative equilibrium of the two gets disturbed in the case of addiction, as impulsive behaviors begin to dominate. As a result, reflective behaviors have little impact on addicts' behavior. One change that ensues is the intensification of addicts' tendency to approach the substances or activities to which they are addicted. This phenomenon is termed approach bias, and plays a key role in addiction to substances such as alcohol. One method that utilizes the above phenomena to alter addicts' behavior is the approach-avoidance test—the first version of which was developed in the 1960s by Solarz [32]. Traditionally, the behavioral component was measured rather rudimentarily by the deflection of a lever (similar to a joystick) away from and toward itself. Such limitations do not exist in VR, where users' behavior can be measured on the approach-avoidance scale more naturally and with greater precision. Research confirming the applicability of this method in VR was conducted by Schroeder et al. [33], who demonstrated that obese people grasped food faster than they grasped neutral stimuli.

Approach-avoidance training programs (AATPs) are effective in supporting standard treatment for alcohol dependence. In 2020, a study on AATPs in VR was announced as a study protocol [34] in BMC Psychiatry. Its main goal is to verify whether AATP training conducted in an environment that simulates a real scenario (a bar created in VR) is more effective than that conducted on a typical computer with the use of a joystick. Due to the article's analyses being limited to the study protocol, its results remain unknown; based on reports on other types of addiction, such as those involving tobacco [35], however, it may be presumed that the method has a chance of proving effective.

In summary, due to its high ecological validity, VR is becoming increasingly useful for clinicians. The technology allows for the presentation and tight control of stimuli from real-life situations, and represents a significant advance in the individualization of AUD assessment and treatment options. The opportunities offered by VR in prolonged exposure improve habituation, which reduces the desire for alcohol-related stimuli and cravings for alcohol itself. Furthermore, multiple exposures under controlled conditions enables patients to build strategies to deal with similar situations in real life. VR techniques rely on the same principles as traditional cognitive-behavioral therapy, but, owing to their ecological validity, enable the potential for the therapy to be applied in patients' lives. Such possibilities are provided by the control of exposure parameters that can be achieved in virtual simulations, and the high degree of similarity between the learning context and real situations in which patients encounter stimuli that induce them to drink alcohol. It appears that techniques that utilize VR can contribute to advancing methods of AUD diagnosis and treatment. They are also applicable to other disorders, such as social anxiety [36, 37] and eating disorders [38].

7.3 Learning Movements in Virtual Reality

VR facilitates simulations of the real world while accurately registering users' movement; it might be presumed, therefore, that the technology could serve as a practical tool in the training of motor tasks. Questionnaires conducted with athletes—a group for which such applications would seemingly be of great use—indicate this to be true—at least on a declaratory basis [39]. The surveyed athletes much more often expected that the use of VR systems would enhance their performance than hinder it. Experimental studies on how the training of movement tasks in VR translates into real-world tasks have been conducted using various technologies, including CAVE and HMDs. Using Kinect sensors, CAVE has been applied to ballet training [40, 41]. Users' movements could be tracked, and the system was able to discern whether they were being performed as the users had been shown on the CAVE system. A similar system for golf learning was developed by Ikeda et al. [42]. By presenting visual feedback on the floor, the system enabled better golf swings. The virtual CAVE system has also been tested as a training tool in American football [43]. Using the SIDEKIQ simulator designed at the University of Michigan, players could experience realistic scenarios during strategy and gameplay training sessions. Initial assessments suggest that the solution is effective, although there are no systematic comparisons between it and the classic training process. Tactical training sessions with the use of VR are also being tested in other sports, including basketball [44, 45]. Studies have shown that VR technology enhances the training process, and that basketball players have no major problems interacting with virtual environments. The effectiveness of VR as a tool for training motor skills has also been tested with the use of modern consumer HMDs, such as Oculus Rift. Uzanowa et al. [46] designed a system for teaching Bulgarian folk dances. While the system did not use additional trackers, it did utilize Kinect and Leap Motion sensors to collect information on the movement of the limbs.

Some reports on VR training sessions are less convinced by their effectiveness. One question that remains on motor training is to what extent the trained abilities are transferable from virtual environments to the real world. These doubts result from the high reality of the experiences delivered in IVR through the use of HMDs. Studies have shown that the transfer of motor skills depends partially on the external context [47]; thus, it is probable that the virtual and real environments are different enough for users to view each in a different context, which causes disturbances in the transferring of motor skills between the two. Anglin et al. and Juliano et al. [48, 49] attempted to establish the validity of this case. Both studies demonstrated that the transfer of skills learned in VR to the real world is disturbed. It should be noted, however, that the studies tested the sequential visual isometric pinch task—one that is highly specific by nature. Questions remain over how the transfer of skills occurs in other tasks.

The use of VR environments reaches beyond the realm of sport: construction firms are attempting to solve skilled labor shortages by holding in-house training sessions using the technology. This is a response to existing teaching methods, which are time-consuming and ineffective. Having collected information on the real demand from construction companies, Wolf et al. developed a VR application that uses low-cost devices, such as Oculus Go, to train construction workers. The application was well received by practitioners [50].

7.4 Risks of Virtual Reality

Despite the rapid pace of its development in recent years, VR continues to carry risks. Although the technology itself is advanced, some of its properties can lead to negative experiences; one of them is cybersickness, which is a leading factor in negative VR experiences. Its symptoms were first described in the 1990s, following a series of studies on the simulator sickness and motion sickness. It was determined that use of simulators is accompanied by a number of potential negative symptoms, including nausea, blurry vision, dizziness, and concentration problems.

A host of theories attempt to explain cybersickness. According to the postural instability theory [51], negative sensations characteristic for motion sickness are caused in situations in which it is impossible for users to maintain stable and correct body posture, or in which they are forced to maintain non-optimal postures for extended periods. Rest frames theory is based on the assumption that movement is always perceived against the backdrop of stillness; the human brain must assume that certain objects are stationary to be able to perceive the movement of others. To observe the movement of a ball in a room, we must be able to determine that the room, which constitutes the background, is stationary. Cyber sickness is the result of disturbances in the mental model that contains representations of moving and static objects [52]. The eye-movement theory postulates that cybersickness occurs when the mechanisms of image stabilization in the retina, which function normally in the real world, fail [53]. The most widely accepted Sensory Conflict Theory assumes that cybersickness is the result of a conflict between the visual and the vestibular systems. This occurs, for example, when the visual system is providing information on the movement of the body (the image on the retina is moving), and the vestibular system on its absence (no changes in the position of the body). This conflict makes the systems of smooth movement in VR environments extremely unpleasant for users [4]. Similar negative impressions can be caused, for example, by rally driving simulators that lack additional stimuli in the form of moving chairs or vibrations. Such vibrotactile stimuli can alter the perception of movement in VR environments [54].

7.4.1 Factors that Influence Cybersickness

Cybersickness can be caused by a host of different factors, including those related to movement. The probability of cybersickness correlates inversely with the naturalness of movements and the level of control users exercise over them. If movements are natural and under a user's control, he/she will feel the negative symptoms of cybersickness to a lesser extent; conversely, when movement is unnatural (for instance, vection, which is produced when moving in a virtual environment involves the user sitting still), negative reactions can be expected [55]. Another relevant factor is acceleration: scenarios in which visual stimuli are inconsistent with the perceived overloads and changes in balance cause unpleasant feelings—the intensity of which depends on the acceleration [56]. Users' fields of view might also cause cybersickness. Larger fields of view enhance immersion and sense of presence [57], but also cause greater problems with cybersickness [58]. Impaired depth perception might also cause discomfort: HMDs do not allow for the appropriate reproduction of depth that would naturally stimulate the eyes, as images are always displayed on screens mounted at a constant distance from them—a similar discomfort to the one frequently caused by 3D televisions [59]. Additionally, the time spent using VR systems, the sense of control, delay (the time between user actions and system reactions), the absence of static reference points, and other visual factors, such as image rotation around the Z axis, contribute to feelings of cybersickness. [60].

VR environments should be prepared with the utmost care to minimize the factors that cause cybersickness. It has been demonstrated that presence correlates negatively with cybersickness: the greater the negative symptoms, the lower the sense among users of being present in virtual environments [61].

References

1. Eysenck, H.J., Eysenck, S.B.G.: Eysenck Personality Inventory (2016). https://doi.org/10.1037/t02711-000
2. Koppensteiner, M.: Motion cues that make an impression: predicting perceived personality by minimal motion information. J. Exp. Soc. Psychol. **49**, 1137–1143 (2013)
3. Słowiński, P., Zhai, C., Alderisio, F., Salesse, R., Gueugnon, M., Marin, L., Bardy, B.G., di Bernardo, M., Tsaneva-Atanasova, K.: Dynamic similarity promotes interpersonal coordination in joint action. J. R. Soc. Interface. **13** (2016). https://doi.org/10.1098/rsif.2015.1093
4. Bohdanowicz, Z., Kowalski, J., Abramczuk, K., Banerski, G., Cnotkowski, D., Kopacz, A., Kobyliński, P., Zdrodowska, A., Biele, C.: Is a virtual ferrari as good as the real one? Children's initial reactions to virtual reality experiences. In: International Conference on Human Systems Engineering and Design: Future Trends and Applications, pp. 397–403. Springer (2019)
5. Biele, C., Grabowska, A.: Sex differences in perception of emotion intensity in dynamic and static facial expressions. Exp. Brain Res. **171**, 1–6 (2006)
6. Aoki, S., Smith, J.B., Li, H., Yan, X., Igarashi, M., Coulon, P., Wickens, J.R., Ruigrok, T.J., Jin, X.: An open cortico-basal ganglia loop allows limbic control over motor output via the nigrothalamic pathway. Elife. **8** (2019). https://doi.org/10.7554/eLife.49995
7. Kimura, D., Harshman, R.A.: Sex differences in brain organization for verbal and non-verbal functions. Prog. Brain Res. **61**, 423–441 (1984)

8. Maccoby, E.E., Jacklin, C.N.: The Psychology of Sex Differences. Stanford University Press (1978)
9. Bradley, M.M., Codispoti, M., Sabatinelli, D., Lang, P.J.: Emotion and motivation II: sex differences in picture processing. Emotion **1**, 300–319 (2001)
10. Hyde, J.S.: Gender similarities and differences. Annu. Rev. Psychol. **65**, 373–398 (2014)
11. Xu, H., Li, L., Fang, M., Zhang, F.: Movement human actions recognition based on machine learning. Int. J. Online Eng. **14**, 193 (2018)
12. Savva, N., Bianchi-Berthouze, N.: Automatic recognition of affective body movement in a video game scenario. In: Intelligent Technologies for Interactive Entertainment, pp. 149–159. Springer Berlin Heidelberg (2012)
13. Brave, S., Nass, C., Hutchinson, K.: Computers that care: investigating the effects of orientation of emotion exhibited by an embodied computer agent. Int. J. Hum. Comput. Stud. **62**, 161–178 (2005)
14. Brave, S., Nass, C.: Emotion in human--computer interaction. In: Human-Computer Interaction Fundamentals, pp. 53–68. CRC Press Boca Raton, FL, USA (2009)
15. Schubring, D., Kraus, M., Stolz, C., Weiler, N., Keim, D.A., Schupp, H.: Virtual reality potentiates emotion and task effects of alpha/beta brain oscillations. Brain Sci. **10** (2020). https://doi.org/10.3390/brainsci10080537
16. Felnhofer, A., Kothgassner, O.D., Beutl, L., Hlavacs, H., Kryspin-Exner, I.: Is virtual reality made for men only? Exploring gender differences in the sense of presence. In: Proceedings of the International Society on Presence Research (2012).
17. Boone, A.P., Gong, X., Hegarty, M.: Sex differences in navigation strategy and efficiency. Mem. Cognit. **46**, 909–922 (2018)
18. Aranha, R.V., Nakamura, R., Tori, R., Nunes, F.L.S.: Personality traits impacts in virtual reality's user experience (2018). https://doi.org/10.1109/svr.2018.00019
19. Kober, S.E., Neuper, C.: Personality and presence in virtual reality: does their relationship depend on the used presence measure? Int. J. Hum. Comput. Interact. **29**, 13–25 (2013)
20. Michailidis, L., Lucas Barcias, J., Charles, F., He, X., Balaguer-Ballester, E.: Combining personality and physiology to investigate the flow experience in virtual reality games. In: HCI International 2019—Posters, pp. 45–52. Springer International Publishing (2019)
21. Wang, Y., Tree, J.E.F., Walker, M., Neff, M.: Assessing the impact of hand motion on virtual character personality. ACM Trans. Appl. Percept. **13**, 1–23 (2016)
22. Ruhland, K., Zibrek, K., McDonnell, R.: Perception of personality through eye gaze of realistic and cartoon models. In: Proceedings of the ACM SIGGRAPH Symposium on Applied Perception, pp. 19–23 (2015)
23. Arellano, D., Varona, J., Perales, F.J., Bee, N., Janowski, K., André, E.: Influence of head orientation in perception of personality traits in virtual agents. In: The 10th International Conference on Autonomous Agents and Multiagent Systems, vol. 3, pp. 1093–1094 (2011)
24. Hu, X., Zhuang, C., Wang, F., Liu, Y.-J., Im, C.-H., Zhang, D.: fNIRS evidence for recognizably different positive emotions. Front. Hum. Neurosci. **13**, 120 (2019)
25. Bandara, D., Hirshfield, L., Velipasalar, S.: Classification of affect using deep learning on brain blood flow data. J. Near Infrared Spectrosc., JNIRS. **27**, 206–219 (2019)
26. Putze, F., Herff, C., Tremmel, C., Schultz, T., Krusienski, D.J.: Decoding mental workload in virtual environments: a fNIRS study using an immersive n-back task. 2019 41st Annual International Conference of the IEEE Engineering in Medicine and Biology Society (EMBC). IEEE 2019,pp. 3103–3106 (2019)
27. Székely, G.J., Rizzo, M.L., Bakirov, N.K.: Measuring and testing dependence by correlation of distances. Ann. Stat. **35**, 2769–2794 (2007)
28. Kobylinski, P., Pochwatko, G., Biele, C.: VR experience from data science point of view: how to measure inter-subject dependence in visual attention and spatial behavior. In: Intelligent Human Systems Integration 2019, pp. 393–399. Springer International Publishing (2019)
29. Muris, P., Merckelbach, H., Holdrinet, I., Sijsenaar, M.: Behavioral avoidance test (2013). https://doi.org/10.1037/t11195-000

30. Mühlberger, A., Sperber, M., Wieser, M.J., Pauli, P.: A virtual reality behavior avoidance test (VR-BAT) for the assessment of spider phobia. J. Cyberther. Rehabil. **1**, 147–158 (2008)
31. Deutsch, R., Strack, F.: Reflective and impulsive determinants of addictive behavior. https://doi.org/10.4135/9781412976237.n4
32. Solarz, A.K.: Latency of instrumental responses as a function of compatibility with the meaning of eliciting verbal signs. J. Exp. Psychol. **59**, 239–245 (1960)
33. Schroeder, P.A., Lohmann, J., Butz, M.V., Plewnia, C.: Behavioral bias for food reflected in hand movements: a preliminary study with healthy subjects (2016). https://doi.org/10.1089/cyber.2015.0311
34. Mellentin, A.I., Nielsen, A.S., Ascone, L., Wirtz, J., Samochowiec, J., Kucharska-Mazur, J., Schadow, F., Lebiecka, Z., Skoneczny, T., Mistarz, N., Bremer, T., Kühn, S.: A randomized controlled trial of a virtual reality based, approach-avoidance training program for alcohol use disorder: a study protocol. BMC Psychiatry **20**, 340 (2020)
35. Eiler, T.J., Grünewald, A., Machulska, A., Klucken, T., Jahn, K., Niehaves, B., Gethmann, C.F., Brück, R.: A preliminary evaluation of transferring the approach avoidance task into virtual reality (2019). https://doi.org/10.1007/978-3-030-23762-2_14
36. Clemmensen, L., Bouchard, S., Rasmussen, J., Holmberg, T.T., Nielsen, J.H., Jepsen, J.R.M., Lichtenstein, M.B.: STUDY PROTOCOL: EXPOSURE IN VIRTUAL REALITY FOR SOCIAL ANXIETY DISORDER—a randomized controlled superiority trial comparing cognitive behavioral therapy with virtual reality based exposure to cognitive behavioral therapy with in vivo exposure (2020). https://doi.org/10.1186/s12888-020-2453-4
37. Bouchard, S., Dumoulin, S., Robillard, G., Guitard, T., Klinger, É., Forget, H., Loranger, C., Roucaut, F.X.: Virtual reality compared with in vivo exposure in the treatment of social anxiety disorder: a three-arm randomised controlled trial. Br. J. Psychiatry. **210**, 276–283 (2017)
38. Ferrer-García, M., Gutiérrez-Maldonado, J.: The use of virtual reality in the study, assessment, and treatment of body image in eating disorders and nonclinical samples: a review of the literature (2012). https://doi.org/10.1016/j.bodyim.2011.10.001
39. Gradl, S., Eskofier, B.M., Eskofier, D., Mutschler, C., Otto, S.: Virtual and augmented reality in sports: an overview and acceptance study. In: Proceedings of the 2016 ACM International Joint Conference on Pervasive and Ubiquitous Computing: Adjunct, pp. 885–888. Association for Computing Machinery, New York, NY, USA (2016)
40. Muneesawang, P., Khan, N.M., Kyan, M., Elder, R.B., Dong, N., Sun, G., Li, H., Zhong, L., Guan, L.: A machine intelligence approach to virtual ballet training. IEEE Multimed. **22**, 80–92 (2015)
41. Kyan, M., Sun, G., Li, H., Zhong, L., Muneesawang, P., Dong, N., Elder, B., Guan, L.: An approach to ballet dance training through MS Kinect and visualization in a CAVE virtual reality environment (2015). https://doi.org/10.1145/2735951
42. Ikeda, A., Hwang, D.-H., Koike, H.: Real-time visual feedback for golf training using virtual shadow. In: Proceedings of the 2018 ACM International Conference on Interactive Surfaces and Spaces, pp. 445–448. Association for Computing Machinery, New York, NY, USA (2018)
43. Huang, Y., Churches, L., Reilly, B.: A case study on virtual reality american football training. In: Proceedings of the 2015 Virtual Reality International Conference, pp. 1–5. Association for Computing Machinery, New York, NY, USA (2015)
44. Tsai, W.-L., Pan, T.-Y., Hu, M.C.: Feasibility study on virtual reality based basketball tactic training (2020). https://doi.org/10.1109/tvcg.2020.3046326
45. Ma, Z., Wang, F., Liu, S.: Feasibility analysis of VR technology in basketball training. IEEE Access 1 (2020)
46. Uzunova, Z., Chotrov, D., Maleshkov, S.: Virtual reality system for motion capture analysis and visualization for folk dance training. In: Proceedings of the 12th Annual International Conference on Computer Science and Education in Computer Science (CSECS 2016), pp. 1–2. Fulda, Germany (2016)
47. Taylor, J.A., Ivry, R.B.: Context-dependent generalization. Front. Hum. Neurosci. **7**, 171 (2013)
48. Anglin, J., Saldana, D., Schmiesing, A., Liew, S.L.: Transfer of a skilled motor learning task between virtual and conventional environments (2017). https://doi.org/10.1109/vr.2017.7892346

49. Juliano, J.M., Liew, S.-L.: Transfer of motor skill between virtual reality viewed using a head-mounted display and conventional screen environments. J. Neuroeng. Rehabil. **17**, 48 (2020)
50. Wolf, M., Teizer, J., Ruse, J.H.: Case study on mobile virtual reality construction training (2019). https://doi.org/10.22260/isarc2019/0165
51. Riccio, G.E., Stoffregen, T.A.: An ecological theory of motion sickness and postural instability. Ecol. Psychol. **3**, 195–240 (1991)
52. Cao, Z., Jerald, J., Kopper, R.: Visually-induced motion sickness reduction via static and dynamic rest frames. In: 2018 IEEE Conference on Virtual Reality and 3D User Interfaces (VR), pp. 105–112 (2018)
53. Flanagan, M.B., May, J.G., Dobie, T.G.: The role of vection, eye movements and postural instability in the etiology of motion sickness. J. Vestib. Res. **14**, 335–346 (2004)
54. Farkhatdinov, I., Ouarti, N., Hayward, V.: Vibrotactile inputs to the feet can modulate vection. In: 2013 World Haptics Conference (WHC), pp. 677–681 (2013)
55. Bonato, F., Bubka, A., Palmisano, S., Phillip, D., Moreno, G.: Vection change exacerbates simulator sickness in virtual environments. Presence. **17**, 283–292 (2008)
56. Stanney, K.M., Kennedy, R.S., Drexler, J.M.: Cybersickness is not simulator sickness. Proc. Hum. Fact. Ergon. Soc. Annu. Meet. **41**, 1138–1142 (1997)
57. Prothero, J. D: Widening the field-of-view increases the sense of presence in immersive virtual environments. Hum. Int. Technol. Lab Tech Rep. (1995)
58. Duh, H.B.L., Lin, J.W., Kenyon, R.V., Parker, D.E., Furness, T.A.: Effects of field of view on balance in an immersive environment. In: Proceedings IEEE Virtual Reality 2001, pp. 235–240 (2001)
59. Eickelberg, S., Kays, R.: Gaze adaptive convergence in stereo 3D applications. In: 2013 IEEE 3rd International Conference on Consumer Electronics ¿ Berlin (ICCE-Berlin), pp. 1–5 (2013)
60. Jung, J.-Y., Cho, K.-S., Choi, J., Choi, J.: Causes of cyber sickness of VR contents: an experimental study on the viewpoint and movement. J. Korea Content. Assoc. **17**, 200–208 (2017)
61. Weech, S., Kenny, S., Barnett-Cowan, M.: Presence and cybersickness in virtual reality are negatively related: a review. Front. Psychol. **10**, 158 (2019)

Chapter 8
Movement in the Environment

8.1 Introduction

Technologies that allow users' movements to be tracked, such as GPS, have been used since the 1990s. Prior to their implementation, other indirect approaches, such as tracking the circulation of banknotes, had been used to track mobility [1]. It was not until the widespread adoption of smartphones, however, that it became much easier to conduct large-scale analyses of movement, pertaining to distances of hundreds of meters or kilometers. Smartphones offer a number of other sensors, such as accelerometers or magnetometers, which facilitate analysis of users' movements on a smaller scale. Motion can also be detected by external devices, such as passive infrared (PIR) sensors or cameras—the images of which are processed with the use of machine learning. These unique capabilities have attracted the attention of both private and state institutions which have recognized them as a valuable source of information on individuals' lifestyles, habits, and needs.

As a result, there are a host of applications on the smartphone market that allow users to track their own activity. The collection and processing of such data allows developers, in turn, to design effective solutions for car navigation, locating places of interest (POI), augmented reality games, systems supporting physical activity, and security—tracking the movements of children, for instance. Admittedly, GPS navigation existed before the dawn of mobile technologies; nevertheless, significant improvements in the effectiveness of mobile navigation applications could not have occurred without the solutions introduced by "tech giants", such as Google Maps—which, by collecting the information on the locations of large numbers of users, are able to suggest the most optimal routes between places, accounting for current traffic and road conditions, and user preferences.

Although analyses on user mobility are undeniably important for human–computer interaction, these discoveries also contribute to research in virology and urban planning.

C. Biele, *Human Movements in Human-Computer Interaction (HCI)*,
Studies in Computational Intelligence 996,
https://doi.org/10.1007/978-3-030-90004-5_8

8.2 Recognition of User Movement on a Small-Scale

Definition of users' locations is necessary not only in open spaces, but also where GPS technology is inaccessible, including indoor areas. The development of smart-home systems and IoT technology allows us humans to use movement to control smart home systems. Such systems can be used to perform simple actions, like switching the lights on when someone enters a room; or to execute complicated scripts that open gates, set the temperature, or trigger other actions when a user enters his/her home. Motion in such systems can be detected via smartphone GPS systems, after connecting mobile phones or other devices to home Wi-Fi networks, or directly, with the use of PIR sensors [2] and RFID tags [3]. The solutions outlined above are becoming increasingly accessible, and are replacing the tracker-based systems first developed in 1990s to locate users within buildings, such as Active Badge by Olivetti. One recent innovative solution comprises an indoor location system based on the measurement of magnetic field disturbances caused by structural steel elements in buildings [4]. To date, however, the solution has never been commercialized.

8.2.1 Multi-layer Localization Monitoring in Industrial Setting

The precision with which individuals can be located raises ethical, social, and legal doubts—and various solutions have been proposed to address these concerns. The issue is particularly relevant in modern industry. The expansion of factories and increasingly complex production processes require intelligent computer systems to support employees in the performance of their tasks [5]. Such systems must be adaptable to specific users, their locations within factories, their current tasks, and the locations of the tools and machinery necessary to perform them. Precise localization is essential for the operation of context-aware systems and industrial processes. While the locations of machinery, tools, or materials does not raise any serious questions, constant monitoring of the locations of humans may be objected to on grounds of privacy. [6, 7]. In order to address these difficulties, Heinz et al. proposed a prototype [8] of an industrial system that preserves the high quality of its localization without compromising the privacy of its users. The multi-level localization system relies on the combined use of several localization technologies of varying accuracy: the first level uses RFID technology, which is implemented on a Raspberry Pi microcomputer; the second uses the Ubisense UWB system; the third uses a Microsoft Kinect V2 camera, in addition to facial recognition algorithms delivered by the Microsoft Face Basics API. These three types of data are integrated into a central system. Such prototypes demonstrate that the implementation of effective localization systems in industrial environments that do not intrude on the privacy of their users is viable.

8.2.2 Recognition of Movement and User Safety

One of the most useful applications of systems based on PIR motion detection sensors—specifically those connected to cameras—is security. If motion is detected by a sensor, the system records a video, which is then analyzed to detect suspicious objects. The efficiency of such systems reaches as high as 90% [9].

Systems designed to detect users' motions within their homes are also crucial for the operation of those designed to ensure the safety of the elderly. Problems associated with alerting emergency services in the right circumstances could be solved by implementation of a localization-based system that responded to specific events—provided that an accurate, robust, and reliable localization system were available. Moreover, the quality of emergency services' responses to such alerts could improve significantly if they had accurate, location-based information on reported emergencies. Research in this area illustrates that PIR sensors in combination with vibration detectors can be used effectively to detect unusual events related to motion, such as falls in the home [10].

Users' movements can also be detected using machine learning systems. Cameras connected to computer vision systems with the assistance of optical flow detection can serve as an effective tool in motion detection. [11]. Such systems are used to recognize the motion patterns of pedestrians in urban environments by combining methods in image processing and pattern detection [12]. In the future, these systems can facilitate the further automation of surveillance camera systems in public places, which are currently operated by humans.

8.2.3 Recognition of User Movement and Ecology

The automatic analysis of user movement provides the means to protect the natural environment by reducing the energy consumption of buildings. Those who are responsible for switching off lights or devices, both in private houses in and public buildings, frequently forget to do so. The Energy Saving Trust estimates that approximately 9–16% of electricity is consumed by devices in standby mode. As regulations pertaining to the amount of energy consumed by such devices become increasingly restrictive, the European Union is encouraging its inhabitants to switch off unused devices and appliances as often as possible [13]. An effective system of tracking users' presence and movements would significantly reduce the energy waste incurred. Tests of the system were implemented at the CMR Institute of Technology laboratory in Karnataka, India [14]. After the implementation of an automatic shutdown control system based on PIR sensors, the average monthly energy consumption of the laboratory decreased from 168 kWh to 109.2 kWh.

8.2.4 Shared Mobility and Ecology

Analyses pertaining to movement around larger areas and over longer distances (within cities, for instance) suggest that solutions such as shared mobility can have a positive impact on the natural environment. Fagnant and Kockelman [15] demonstrated in their simulations that one shared vehicle could replace 11 conventional ones; the simulations proffered, however, that the distance covered by the vehicles would increase by approximately 10%. The conclusions pertaining to such aspects as the waiting time for transport—which is important for users' convenience and, as a result, for the potential adoption of the system—appear highly promising. It is worth mentioning that the simulations were run before the onset of the COVID-19 pandemic, and did not consider the potential hygiene challenges of vehicle-sharing systems.

8.2.5 Other Uses of Human Movement Detection Algorithms

Methods developed with the detection of human movement in mind are also used in the analysis of animal migration [16, 17]. Machine vision can be employed to recognize animal movements [18], and is able to identify behaviors caused by animal welfare problems. This is of high value to breeders, both ethically and economically.

8.3 Recognition of User Movement Patterns from the Perspective of Urban Mobility

Due to the colossal amount of data generated by moving users, initial analyses in the field were based on aggregation. Alternatively, researchers would consciously sacrifice the accuracy offered by GPS; instead relying on less accurate, but more accessible, data, such as that of GSM signals [19]. By the mid-2000s, Sohn et al. had demonstrated that it was possible to calculate the daily number of steps taken, and to recognize different types of movement based on information from GSM signals. Their system relied on a fundamental property of the radio signal used in GSM technology: that the signal is consistent in time, but variable in space. The researchers demonstrated that the dynamics of changes in the radio signals for different types of movement was identifiable and distinct; for example, they were much faster while travelling by car than on foot. This allowed the authors to identify how cellphone users moved, and to estimate the distance they had covered (in other words, how many steps they had taken in a given day) fairly accurately. Nevertheless, as smartphones and IoT technologies have risen in popularity, localization tracking based on GSM signals has lost much of its relevance.

8.3.1 Localization Based on GPS Data

The significantly more accurate data available from GPS and mobile technologies has facilitated the implementation of solutions based on recorded movement data. A key component of these solutions is mobility clustering—the purpose of which is to identify and group similar trajectories of movement. In this case, algorithms designed to measure distances from assumed trajectories are most commonly used [20], and regression models are occasionally used [21]. Based on the results of studies, researchers have also developed systems to recommend routes [22], or to search for similar users based on their location histories [23].

8.3.2 Predicting User Movement

Another compelling area of the research on localization lies in the prediction of movements. Reports based on the analysis of telephone call histories have indicated that such systems deliver efficiency of more than 93% [24]. It is notable, however, that such efficiency might be related directly to the data analysis methodology, which is based on time bins; and to spatiotemporal resolution, which is used in this type of analysis. Humans frequently remain in a single place for extended periods of time. For that reason, in analyses based on periods shorter than the average time spent in one place, stationarity would result in misleading boosts in the algorithms' effectiveness. Objections of this kind have prompted researchers to seek new approaches to the subject—one of which, developed as part of The Copenhagen Network Study [25], involves simply predicting the next location, rather than the location in the next time bin. These analyses clearly indicate that, after removing the effects of stationarity, the predictability of the next location is significantly lower (around 70%) than that of the location in the next time bin (over 90%).

8.3.3 Combining GPS and External Data

Combining GPS and other public transport network data ensures greater accuracy, and offers more information on how people move [26]. In their Chicago transportation system experiments, Stenneth et al. demonstrated that using a model constructed on the basis of timetables, and bus and bus stop locations, it was possible to achieve 93.5% accuracy when inferring data from various modes of transport, compared to approximately 76% when the data was received solely from GPS. Of all the classifiers used in the experiments (Bayesian Net, Decision Tree, Random Forest, Naive Bayes, and Multi-layer Perceptron), Random Forest was found to be the most efficient. Learning patterns of human behavior from sensor data is essential for the inference of higher-level activity—not only to determine what users are doing (such as walking or driving) and where they are doing it, but also to determine what places are of special interest to those users (such as workplaces, schools, or homes). The attempts to date [27] indicate that this type of inference achievable with the use of hierarchical conditional random fields.

8.4 Navigation and Human Behavior

Cognitively, the navigation process can be divided into two key sub-processes: wayfinding (planning how to reach destinations), and locomotion (the implementation of those plans). Moving in an environment is a crucial skill for humans. Before GPS became widespread, navigation required knowledge on the properties of an environment, such as curves or views along routes [28]; presently, however, when almost every cellphone is fitted with a GPS receiver and navigation software (such as Waze or Google Maps), the requirements for individuals on the move, especially in unknown environments, have altered significantly. Navigation systems offer simple instructions on where to go. When such systems became more advanced, however, some began to highlight their potential negative impact on spatial knowledge acquisition [29]. Although navigation systems decrease the cognitive effort required to accomplish relocation tasks, they do not necessarily induce an absolute reduction, as navigating with an external device might require its user to be more focused, and to be ready to operate the device. In the best case scenario, use of such devices results in users receiving limited information on the physical environment through which they are traveling; in the worst case, it might even lead to death [30]. Analyses of reports on dangerous road incidents (Table 8.1) conducted by Lin et al. exhibit that most incidents related to the use of navigation involve collisions with stationary objects or collisions with other vehicles. These are often caused by distraction connected with using navigation devices. Importantly, in incidents caused by the use of navigation devices, the greater the role of the user's distraction in a given case, the more serious its effects were.

Table 8.1 Incident types connected with navigation use (adapted from [30])

Types of incident	%
Trespass (violate space usage rules)	3
Wrong way (opposite side)	4
Detour (e.g. wrong address)	16
Stranded/stuck (e.g. in the wildness, on railroad tracks)	20
Crashes	57
Crashes with pedestrians/bikes	8
Crashes with vehicles	17
Single-vehicle collisions	32

8.5 Spatial Crowdsourcing

The term, "crowdsourcing" was coined by Jeff Howe in an article published in Wired magazine in 2006, and defines a paradigm in which individual users are engaged to perform tasks that were once performed within companies and institutions. It is particularly suited to tasks that are inherently easier for humans than for computers. Online crowdsourcing easily allows a large number of people to be involved when that is necessary for a task to be performed effectively. The approach has developed over the years, and as location monitoring technology and the so-called "sharing economy" evolved, it was hypothesized that crowdsourcing could also prove useful in the performance of tasks that are spatial and temporal in nature. The use of spatial crowdsourcing increases human potential to perform tasks pertaining to "real" situations involving physical locations, which is unfeasible using typical crowdsourcing solutions. The most distinctive feature of spatial crowdsourcing is the presence of spatial tasks whose completion requires workers to be physically present at a specific location. Solutions that go beyond online-only tasks have recently appeared on the market; platforms like Uber and Gigwalk employ individuals to perform tasks that require movement and physical presence at specific locations at specific times. The concept of spatial crowdsourcing [31] builds on the existing phenomenon of participatory sensing, which focuses primarily on the transmission of information from sensors (such as those fitted in cellphones) by users.

8.5.1 Psychological Factors Influencing Spatial Crowdsourcing

From the perspectives of human-technology interaction and human–computer interaction, it is useful to identify what factors motivate people to participate in spatial crowdsourcing initiatives. After all, such activities require significantly more commitment than the crowdsourcing tasks that are typically performed on computers. Large platforms, such as Amazon Mechanical Turk, focus chiefly on simple, repetitive tasks for which their users are poorly paid. Services like TaskRabbit, on which workers can perform home repair services in their local areas, specialize in non-repetitive and more challenging, but better paid tasks. The table below shows popular tasks commissioned on the most popular platforms.

Platform	Typical tasks
Field Agent	Price checks, service assessment, photography, verification, property evaluation
Gigwalk	Data collection, photography, focus Groups, store audits, IT, financial services
NeighborFavor	Deliveries, rides, groceries

(continued)

(continued)

Platform	Typical tasks
TaskRabbit	Groceries, moving, deliveries, cleaning
WeGoLook	Verifying properties, automobiles, dates, boats, heavy equipment before purchase

Having analyzed the division of tasks among users of one spatial crowdsourcing platform, Musthag and Ganesan [32] found that 10% of the most active users performed over 80% of the tasks, and earned over 80% of all the payments. This means that only highly motivated people are able to receive substantial remuneration from such platforms. It is worth exploring why so many tasks are commissioned to so-called "super-users". These users use strategies that allow them to achieve such high efficiency—one of which involves grouping tasks by location; more than 50% of the super-users' income derived from sessions during which they had performed at least two tasks. Average users combined 5% of their tasks, while super-users did so four times more often. This implies that spatial crowdsourcing systems will have to face the challenge of optimizing routes. At present, users do that themselves. Providing that sort of functionality would undoubtedly contribute to an increase in the number of users and, in consequence, to the effectiveness of such platforms. The high efficiency of super-users can also be explained by their tendency to select better paid tasks, and their speed of task completion—in the case of some tasks, even up to three times faster than the average user.

8.5.2 Other Crowdsourcing Paradigms Involving Movement

Given how rapidly technology continues to develop, and how prominent its presence in our lives has become, much attention is being paid to new solutions in which user movement is a key component. The process of crowdsourcing development and the shift from online-only to physical tasks have led to industry developments such as spatial crowdsourcing, which can be placed in the same category, but differ in their details:

- **Crowdsensing**

Crowdsensing, which is also referred to as "participatory sensing" or "social sensing", utilizes in-built smartphone sensors to collect environmental information, such as locations or temperatures. It is based on user mobility and relies on portable Wi-Fi network data [33]. Crowdsensing involves passively collecting data from sensors in smartphones, and sending it to a server for further processing and analysis.

- **Situated crowdsourcing**

This technique utilizes displays, such as tablets or touch screens, to collect data from individuals who are physically present at a specific location. Such systems are used, for example, to generate ideas, solve problems or gather opinions [34, 35].

8.6 Location Monitoring and Privacy

The examples above demonstrate that methods for locating users in spaces, both large and small-scale, have developed significantly in recent years. Presently, smartphone ownership is widespread; location-tracking systems run at all times, and store information on the places users have spent their time—from specific aisles in libraries, to the monuments visited during last year's holiday. This catalogue of data might include highly sensitive and private information pertaining to users' health, for instance, which can be inferred from the locations they have visited, such as medical centers. As IoT technology matures, there is certainly going to be more user movement tracking data in circulation. Services that monitor the locations and movements of their users are now known as location-based services.

The earliest indoor location systems, such as Active Badge, assumed, dangerously, that all users were trustworthy, and information on their locations remained accessible to all other users [36]. This vulnerability was addressed fairly rapidly, and privacy issues in newer systems have been considered at the design stage and throughout the development process [37].

Location-related privacy issues have been discussed since the development of technology enabling the large-scale tracking of user movement. Various solutions have attempted to mitigate the threats—ranging from those in which users decide on the accuracy of shared locations to those based, for instance, on sharing rules [38]. A classic example can be found in the case of mix-zones [39], in which, analogous to the approach adopted in networks, the anonymity of traffic is ensured by appropriate mixing of data from many senders. Data related to the identity of individuals in specific locations is typically mixed. In recent years, a host of other technological solutions have appeared in the literature to protect users' location data, including cloaking, obfuscation, and caching.

8.6.1 User Perspective on Location Privacy

The results of user perspective studies [40] illustrate that 78% of smartphone users are aware of the risk of applications having access to their locations; 85% stated that it was important to know who had access to such data. Due to the potential inaccuracy of users' declarations, particularly with regard to privacy issues, an interesting method of determining the actual value of privacy for users involves expressing it in monetary terms [41]. In a study that fused psychology and economics, Cambridge students were invited to participate in a (fictitious) study, in which they were requested to share their locations for 28 days. The median bid proposed by the subjects as compensation for participating in the study was 10 GBP—equivalent to less than 0.5% of the monthly average remuneration in the United Kingdom at the time of the study. The authors describe this sum as neither trivial nor particularly large. Interestingly, the sum demanded by the subjects doubled when informed that their location data would be shared with external commercial entities.

8.6.2 Privacy and Shared Mobility

Another area in which user privacy can be compromised is shared transport. The term can be defined as any vehicle-sharing solution in which travelers share a vehicle (either as a group (e.g. ride-sharing), or for a fixed period, (e.g. carsharing). Such solutions allow users to share travel costs and to choose between public and private transport, depending on their travel needs. It encompasses a wide variety of modes of transport, although cars, bicycles, and scooters are by far the most frequently used. Shared mobility has been recognized to contribute to reductions in traffic and congestion; and, in consequence, to less demand for parking spaces, lower transportation costs in urban and suburban areas, and reductions in carbon dioxide emissions. It might also strengthen social relationships [42]. Shared transport owes its success to advanced location solutions, implemented both within the means of transport and among the users themselves. In sharing their locations with systems, users are vulnerable to exposing their behaviors or personal habits. Simultaneously, the development of shared transport solutions encourages the delineation between technology and transport to blur. There are also an increasing number of data leaks from large websites.

It is questionable whether privacy could be protected in the case of products that are so inherently related to the locations of their users, such as shared transport services; and, if so, whether the protection of privacy (achieved, for instance, through location anonymization) would entail a reduction in the quality of those services. Although concealing data protects users' privacy, the more accurate the data in the system is, the higher the quality of the service is. Research on the trade-off between privacy and the cost and quality of transport services was conducted in 2020 by Martelli et al. at MIT JTL Mobility Labs [43]. Their studies demonstrate that the implementation of increased protection for user locations leads both to decreased efficiency of systems and reductions in the quality of services on offer.

It seems that although researchers recognize the importance of privacy in location services, and privacy remains crucial for users, too few solutions are currently implemented in the form of products that are accessible to users.

8.6.3 *Privacy and Spatial Crowdsourcing*

As in the case of shared mobility, user privacy remains a challenge in spatial crowdsourcing. Typically, workers are required to share their locations before selecting tasks, or to undergo verification processes related to their performance.

Knowing the location of users, attackers are able to infer sensitive data, including health information, based on their visits to specific medical centers; religious preferences, based on their visits to specific temples; and lifestyle preferences, based on their visits to specific entertainment venues. It is impossible to fully protect any individual's privacy, even if a false identity is used. Location data can be used to identify patterns of movement, which can lead to exposure of information on users' workplaces or places of residence [44], and, in consequence, of their identities.

The prominent approaches toward privacy in spatial crowdsourcing include:

- Pseudonymization techniques: separating users' identities from the data they send [45];
- Cloaking techniques: concealing the precise locations of workers within a cloaked region [46];
- Exchange-based techniques: initially, information is exchanged directly between system users; in the next stages, transaction details containing no sensitive data are sent to servers [47];
- Encryption-based techniques: concealing the locations and identities of users [48];
- Differential privacy-based techniques: concealing the locations of users by distorting location data with artificial noise [49, 50].

Most of the current spatial crowdsourcing literature focuses on protecting the privacy of workers' locations, while that of the users who outsource the tasks is rarely discussed; the locations of the tasks presented on spatial crowdsourcing platforms is public. With this in mind, solutions are being developed to protect the privacy of task requesters; for example, Liu et al. [51] propose solutions in which the locations of tasks and of the individuals who commission them are protected by an encryption-based mechanism. Protection can also be ensured by perturbation-based mechanisms [52].

While the subject of user motivation in classical crowdsourcing has been discussed widely, studies on spatial crowdsourcing remain sparse. Assumedly, much of the knowledge on traditional crowdsourcing would also be applicable to some, but not all, types of task performed in the "real" world. One of the few studies on the subject [53] concludes that level of trust is a motivational factor for female users of spatial crowdsourcing platforms that enable performance of physical tasks to be requested online.

8.7 Conclusions

Tremendous progress has been made in the area of human location monitoring, since the research using banknotes. In the modern world, information systems related to the movement of people in cities or buildings, such as modern factories, play a large role. At the same time, the growing popularity of such systems raises questions about privacy issues. Systems such as spatial crowdsourcing or even a seemingly ordinary navigation application running on a smartphone require constant sharing of information about one's location. At the same time, systems that track movement in buildings can be used to increase the safety of, for example, the elderly.

References

1. Brockmann, D., Hufnagel, L., Geisel, T.: The scaling laws of human travel. Nature **439**, 462–465 (2006)
2. Yang, D., Xu, B., Rao, K., Sheng, W.: Passive infrared (PIR)-based indoor position tracking for smart homes using accessibility maps and A-star algorithm. Sensors **18** (2018). https://doi.org/10.3390/s18020332
3. Jiang, X., Liu, Y., Wang, X.: An enhanced approach of indoor location sensing using active RFID. In: 2009 WASE International Conference on Information Engineering, pp. 169–172 (2009)
4. Chung, J., Donahoe, M., Schmandt, C., Kim, I.J., Razavai, P., Wiseman, M.: Indoor location sensing using geo-magnetism. In: Proceedings of the 9th International Conference on Mobile Systems, Applications, and Services, pp. 141–154. Association for Computing Machinery, New York, NY, USA (2011)
5. Fellmann, M., Robert, S., Büttner, S., Mucha, H., Röcker, C.: Towards a framework for assistance systems to support work processes in smart factories. In: Machine Learning and Knowledge Extraction, pp. 59–68. Springer International Publishing (2017)
6. Röcker, C., Feith, A.: Revisiting privacy in smart spaces: social and architectural aspects of privacy in technology-enhanced environments. In: Proceedings of the International Symposium on Computing, Communication and Control, pp. 201–205. Citeseer (2009)
7. Sack, O., Röcker, C.: Privacy and security in technology-enhanced environments: exploring users' knowledge about technological processes of diverse user groups. Univers. J. Psychol. **1**, 72–83 (2013)
8. Heinz, M., Büttner, S., Wegerich, M., Marek, F., Röcker, C.: A Multi-level localization system for intelligent user interfaces. In: Distributed, Ambient and Pervasive Interactions: Technologies and Contexts, pp. 38–47. Springer International Publishing (2018)
9. Surantha, N., Wicaksono, W.R.: Design of smart home security system using object recognition and PIR sensor. Proced. Comput. Sci. **135**, 465–472 (2018)
10. Yazar, A., Cetin, A.E.: Ambient assisted smart home design using vibration and PIR sensors (2013). https://doi.org/10.1109/siu.2013.6531531
11. Xu, H., Li, L., Fang, M., Zhang, F.: Movement human actions recognition based on machine learning. Int. J. Online Eng. **14**, 193 (2018)
12. Chavat, J., Nesmachnow, S., Tchernykh, A., Shepelev, V.: Active safety system for urban environments with detecting harmful pedestrian movement patterns using computational intelligence. NATO Adv. Sci. Inst. Ser. E Appl. Sci. **10** 9021 (2020)
13. Your electricity is dripping away. Stop it. Consumer guide on standby losses of appliances, https://ec.europa.eu/energy/intelligent/projects/sites/iee-projects/files/projects/documents/selina_consumer_guide_en.pdf

14. Harsha, B.K., N., N.K.G.: Home Automated power saving system using PIR sensor (2020). https://doi.org/10.1109/icirca48905.2020.9183050
15. Fagnant, D.J., Kockelman, K.M.: The travel and environmental implications of shared autonomous vehicles, using agent-based model scenarios. Transp. Res. Part C: Emerg. Technol. **40**, 1–13 (2014)
16. Patterson, T.A., Basson, M., Bravington, M.V., Gunn, J.S.: Classifying movement behaviour in relation to environmental conditions using hidden Markov models. J. Anim. Ecol. **78**, 1113–1123 (2009)
17. Demšar, U., Buchin, K., Cagnacci, F., Safi, K., Speckmann, B., Van de Weghe, N., Weiskopf, D., Weibel, R.: Analysis and visualisation of movement: an interdisciplinary review. Mov. Ecol. **3**, 5 (2015)
18. Tscharke, M., Banhazi, T.M.: A brief review of the application of machine vision in livestock behaviour analysis. Agrárinformatika/J. Agric. Inform. **7**, 23–42 (2016)
19. Sohn, T., Varshavsky, A., LaMarca, A., Chen, M.Y., Choudhury, T., Smith, I., Consolvo, S., Hightower, J., Griswold, W.G., de Lara, E.: Mobility detection using everyday gsm traces. In: UbiComp 2006: Ubiquitous Computing, pp. 212–224. Springer Berlin Heidelberg (2006)
20. Pelekis, N., Kopanakis, I., Ntoutsi, I., Marketos, G., Theodoridis, Y.: Mining trajectory databases via a suite of distance operators (2007). https://doi.org/10.1109/icdew.2007.4401043
21. Gaffney, S., Smyth, P.: Trajectory clustering with mixtures of regression models. In: Proceedings of the 5th ACM SIGKDD International Conference on Knowledge Discovery And Data Mining, pp. 63–72. Association for Computing Machinery, New York, NY, USA (1999)
22. Yoon, H., Zheng, Y., Xie, X., Woo, W.: Smart Itinerary Recommendation Based on User-Generated GPS Trajectories. In: Ubiquitous Intelligence and Computing, pp. 19–34. Springer Berlin Heidelberg (2010)
23. Xiao, X., Zheng, Y., Luo, Q., Xie, X.: Finding similar users using category-based location history. In: Proceedings of the 18th SIGSPATIAL International Conference on Advances in Geographic Information Systems, pp. 442–445. Association for Computing Machinery, New York, NY, USA (2010)
24. Song, C., Qu, Z., Blumm, N., Barabási, A.-L.: Limits of predictability in human mobility. Science **327**, 1018–1021 (2010)
25. Ikanovic, E.L., Mollgaard, A.: An alternative approach to the limits of predictability in human mobility. EPJ Data Sci. **6**, 12 (2017)
26. Stenneth, L., Wolfson, O., Yu, P.S., Xu, B.: Transportation mode detection using mobile phones and GIS information. In: Proceedings of the 19th ACM SIGSPATIAL International Conference on Advances in Geographic Information Systems, pp. 54–63. Association for Computing Machinery, New York, NY, USA (2011)
27. Liao, L., Fox, D., Kautz, H.: Extracting places and activities from GPS traces using hierarchical conditional random fields. Int. J. Rob. Res. **26**, 119–134 (2007)
28. Downs, R.M., Stea, D.: Image and Environment: Cognitive Mapping and Spatial Behavior. Transaction Publishers (2017).
29. Gardony, A.L., Brunyé, T.T., Mahoney, C.R., Taylor, H.A.: How navigational aids impair spatial memory: evidence for divided attention. Spat. Cogn. Comput. **13**, 319–350 (2013)
30. Lin, A.Y., Kuehl, K., Schöning, J., Hecht, B.: Understanding "death by GPS": a systematic analysis of catastrophic incidents associated with personal navigation technologies. CHI 2017 (2017)
31. Kazemi, L., Shahabi, C.: GeoCrowd: enabling query answering with spatial crowdsourcing. In: Proceedings of the 20th International Conference on Advances in Geographic Information Systems, pp. 189–198. Association for Computing Machinery, New York, NY, USA (2012)
32. Musthag, M., Ganesan, D.: Labor dynamics in a mobile micro-task market. In: Proceedings of the SIGCHI Conference on Human Factors in Computing Systems, pp. 641–650. Association for Computing Machinery, New York, NY, USA (2013)
33. Zenonos, A., Stein, S., Jennings, N.: An algorithm to coordinate measurements using stochastic human mobility patterns in large-scale participatory sensing settings. AAAI, vol. 30 (2016)

34. Goncalves, J., Ferreira, D., Hosio, S., Liu, Y., Rogstadius, J., Kukka, H., Kostakos, V.: Crowdsourcing on the spot: altruistic use of public displays, feasibility, performance, and behaviours. In: Proceedings of the 2013 ACM International Joint Conference On Pervasive and Ubiquitous Computing, pp. 753–762. Association for Computing Machinery, New York, NY, USA (2013)
35. Hosio, S., Goncalves, J., Kostakos, V., Riekki, J.: Crowdsourcing public opinion using urban pervasive technologies: lessons from real-life experiments in Oulu. Policy Internet **7**, 1–20 (2015)
36. Want, R., Hopper, A., Falcão, V., Gibbons, J.: The active badge location system. ACM Trans. Inf. Syst. Secur. **10**, 91–102 (1992)
37. Priyantha, N.B., Chakraborty, A., Balakrishnan, H.: The Cricket location-support system. In: Proceedings of the 6th Annual International Conference on Mobile Computing and Networking, pp. 32–43. Association for Computing Machinery, New York, NY, USA (2000)
38. Hengartner, U., Steenkiste, P.: Protecting access to people location information. In: Security in Pervasive Computing, pp. 25–38. Springer Berlin Heidelberg (2004)
39. Beresford, A.R., Stajano, F.: Location privacy in pervasive computing. IEEE Pervasive Comput. **2**, 46–55 (2003)
40. Fawaz, K., Shin, K.G.: Location privacy protection for smartphone users. In: Proceedings of the 2014 ACM SIGSAC Conference on Computer and Communications Security, pp. 239–250. Association for Computing Machinery, New York, NY, USA (2014)
41. Danezis, G., Lewis, S., Anderson, R.J.: How much is location privacy worth? In: WEIS. Citeseer (2005)
42. Librino, F., Renda, M.E., Santi, P., Martelli, F., Resta, G., Duarte, F., Ratti, C., Zhao, J.: Home-work carpooling for social mixing. Transportation **47**, 2671–2701 (2020)
43. Martelli, F., Renda, M.E., Zhao, J.: The price of privacy control in mobility sharing. J. Urb. Technol. 1–26 (2020)
44. Huang, C., Wang, D., Zhu, S.: Where are you from: home location profiling of crowd sensors from noisy and sparse crowdsourcing data. In: IEEE INFOCOM 2017—IEEE Conference on Computer Communications, pp. 1–9 (2017)
45. Cornelius, C., Kapadia, A., Kotz, D., Peebles, D., Shin, M., Triandopoulos, N.: Anonysense: privacy-aware people-centric sensing. In: Proceedings of the 6th International Conference on Mobile Systems, Applications, and Services, pp. 211–224. Association for Computing Machinery, New York, NY, USA (2008)
46. Kazemi, L., Shahabi, C.: A privacy-aware framework for participatory sensing. ACM Sigkdd Explor. Newsl. (2011)
47. Zhang, B., Liu, C.H., Lu, J., Song, Z., Ren, Z., Ma, J., Wang, W.: Privacy-preserving QoI-aware participant coordination for mobile crowdsourcing. Comput. Netw. **101**, 29–41 (2016)
48. Shen, Y., Huang, L., Li, L., Lu, X., Wang, S., Yang, W.: Towards preserving worker location privacy in spatial crowdsourcing. In: 2015 IEEE Global Communications Conference (GLOBECOM), pp. 1–6 (2015)
49. To, H., Ghinita, G., Shahabi, C.: A framework for protecting worker location privacy in spatial crowdsourcing. Proc. VLDB Endow. **7**, 919–930 (2014)
50. Dai, J., Qiao, K.: A privacy preserving framework for worker's location in spatial crowdsourcing based on local differential privacy (2018). https://doi.org/10.3390/fi10060053
51. Liu, A., Wang, W., Shang, S., Li, Q., Zhang, X.: Efficient task assignment in spatial crowdsourcing with worker and task privacy protection. GeoInformatica **22**, 335–362 (2018)
52. To, H., Shahabi, C., Xiong, L.: Privacy-preserving online task assignment in spatial crowdsourcing with untrusted server. In: 2018 IEEE 34th International Conference on Data Engineering (ICDE), pp. 833–844 (2018)
53. Mahmod, M., Hassan, H.: Spatial crowdsourcing: Opportunities and challenges in motivating Malaysian women's participation in Gig economy. In: 2020 IEEE Conference on e-Learning, e-Management and e-Services (IC3e), pp. 76–81. IEEE (2020)

Chapter 9
Perception of Movement

9.1 Introduction

When interacting with computers or smartphones, most of the stimuli displayed on their screens remain static. During specific types of interaction, however, motion becomes a crucial element. It is also used as an element of the interfaces that enhance user experience and support interaction. In the case of interactions with avatars—for instance, in VR—motion is a key element. In order to better understand the problem of perception of movement during human–computer interaction, the subject should be approached from a general psychological and neurobiological perspective.

The human brain displays unique capabilities related to the perception of movement—especially that of biological origin. Any movement we observe is automatically reflected in our brains, and this allows us to understand our environment. Movement is also an inherent element of human–computer interaction in the form of either moving elements of the interface, or of humanoid avatars controlled by AI. Designers of both traditional and newly-emerging interfaces that utilize virtual or augmented reality, therefore, would be well-advised to draw from the extensive psychological and neurobiological knowledge on the human perception of movement.

9.1.1 What is Perception?

Perception is a complex cognitive process that leads to the formation of an image of reality in an individual's mind. Perception is not merely a passive reflection of reality, but an active, creative process that involves compiling the material provided by each of the senses into a whole. The information received from sense organs is processed on two levels: sensory and motor, and semantic and operational. In an active process of perception, the pieces of information an individual has already possessed—such as his/her attitudes and cognitive schemata—play a significant role. These are treated

C. Biele, *Human Movements in Human-Computer Interaction (HCI)*,
Studies in Computational Intelligence 996,
https://doi.org/10.1007/978-3-030-90004-5_9

as hypotheses that are verified by the data from the environment. Since the data already possessed plays such an integral role in perception, it remains possible to correctly recognize objects or phenomena in spite of gaps in the sensory material, and in spite of its variability resulting from the conditions in which a process is occurring. It is worth mentioning at this point that the influence of the senses varies when constructing perceptual impressions: sight is a dominant sense. This has been proven by studies involving enologists—experts in the science of wine. In order to describe a wine's qualities, enologists use separate vocabularies reserved for white and red wines. When, as part of an experiment, they were asked to evaluate white wine dyed red, they continued to use the terms assigned to evaluating red wines [1]. The impact of visual stimuli on judgment becomes even more apparent when movement stimuli are perceived. This has been proven in studies that involved evaluation of musical performances presented in the form of video clips [2]; in one version of the recordings the musicians remained still, while in the other they moved expressively. The scores allocated by others watching the recorded performances were higher in the case of the moving musicians. This suggests that perceived movement can affect judgment of the properties of objects that are not directly accessible. Such objects might include broadly defined computer systems—the evaluation of which can be manipulated by appropriate implementation of moving visual stimuli, or by implementing movement itself, such as in the case of robots.

9.2 Perception of Movement

The impression of movement is created when the objects perceived move around the surface of the eye's retina. If, from moment to moment, the same image appears in a slightly different part of the retina, the human brain interprets this as movement of the perceived object. Additionally, the brain features mechanisms that protect us against perception errors that might exist in situations when we move our eyes—this occurs constantly when we are looking at something. In such circumstances, the retinal image of perceived objects changes, yet the objects remain stationary. These mechanisms are possible due to signals that originate in the eye muscles.

In the processes of perception of movement, there are two special categories of stimuli: a) biological movement concerns moving, animate objects, such as human or animal silhouettes; b) implied motion concerns static "photographs" of moving objects, such as still images of flying missiles or jumping men. Perception of this type of stimuli occurs frequently in the case of human–computer interactions.

9.2.1 Biological Movement Perception

One of many interesting examples of the unique capabilities of the human visual system is the phenomenon of perception of biological movement [3]—of living organisms, and particularly of humans. Initial studies of the subject utilized the technology of point-light displays [4], which involves placing light points on the head and joints—including the shoulders, elbows, and knees—of a person filmed in the dark.

Animations created with point-light technology are able to render many different types of action in a manner that is easy to recognize using approximately 10–13 points in most cases—none of which contain any information on the action, nor on the person moving. Unsurprisingly, single frames extracted from this type of animation present a chaotic and unrecognizable arrangement of points. It is only when they are presented in motion that the points combine into a whole, allowing viewers to quickly recognize the human form, and to distinctly identify the action performed, such as dancing or weight lifting [5]. This suggests that perception of moving living organisms requires universal integration, in time and space, of information regarding movement. This integration occurs with great efficiency; based on observations of biological movement it is possible to discern, among other things, an individual's sex [6], identity [7], or emotional condition [8, 9]. Point-light display studies have also been conducted with respect to the processes of perceiving expressions of emotion. A set of pioneering experiments conducted in the 1970s by John Bassili remain, however, the rare contribution to the field [1, 2]. In his studies, participants were shown both moving and static expressions created using point-light display technology. The participants were capable of correctly recognizing emotional expressions when they were presented as animations. Bassili proved that information about the configuration of individual parts of the face were unnecessary for the cognitive processes of recognizing expressions of emotion to occur—the perception of movement was adequate.

In the present day, studies involving advanced motion-capture systems have, by and large, superseded video solutions based on point-light displays [3–5].

9.2.2 The Neural Basis of Perception of Biological Movement

Despite a large amount of data concerning the organization of the visual system coming from anatomical studies conducted on primates, it is believed that the observed separation into functionally distinguishable pathways also applies to humans. Based on current knowledge, it is believed that visual information is transmitted inside the brain over two pathways: the ventral, which is responsible for processing information about form; and the dorsal, which processes movement and the visual co-ordination of actions aimed at objects [6, 7].

Martin Giese and Tomaso Poggio [8] proposed a model of biological movement perception that integrated the results of existing psychophysical and neuropsychological studies. The model has a hierarchic structure—in successive areas, the analysis of a perceived object occurs at an increasingly higher level of generality, and in an increasingly large receptive field. The model is also compatible with the existence of functionally different pathways. The role of these pathways in perception of biological movement is unclear, but it seems that both play a fundamental role in the process. The significance of the dorsal pathway is evidenced by the results, which indicate that it is possible to perceive actions without simultaneously perceiving information concerning the object's shape [9, 19]. Humans are capable, however, of recognizing gait patterns from static key frames [10], or stimuli with disrupted information about movement [11]. Neural imaging studies have also confirmed the existence of neurons responsible for perceiving the shape of the human body [12, 13]. It is therefore likely that biological movement is recognized on the basis of perception of sequences of static images of body shapes [8], which would suggest the involvement of the ventral pathways in this process. It is noteworthy that the two pathways are not entirely separate; information transmitted over both the ventral and the dorsal pathways arrives in the superior temporal polysensory area (STPa). There are reasons to suspect that information on movement and shape is integrated in this area, and that it is responsible for the perception of biological movement [14].

In neural imaging studies conducted on humans, scientists have confirmed activation of various cortical and subcortical areas during the perception of biological movement. Howard et al. [15] detected bilateral activation in both the medial temporal lobe (MTL), and in the adjacent superior temporal gyrus (STG). In other studies, activations occurred in the superior temporal sulcus (STS) and the amygdalae, which are the areas involved in recognizing emotional expressions. Conversely, no excitations in the MTL were detected [16]. In yet other studies, activations appeared both in the STS and in the MTL when perceiving biological movement. Stimulations observed in the area of the STS are consistent with the aforementioned role of the STPa in the perception of movement, as the STS and the STPa are considered to be areas analogous in both humans and apes [17].

Data concerning the locations of the functions associated with the perception of biological movement has also been provided by clinical studies. It has been demonstrated that patients with damaged occipital-temporal and parietal-temporal areas have no problems with perceiving biological movement, despite suffering multiple disorders of basic visual functions [18, 19]. These patients had suffered damage to the MTL and the adjacent areas responsible for processing movement; as a result, the STS failed to receive information from the dorsal pathway. The patients retaining their capacity to perceive movement suggests the existence of another path that specializes in perceiving motion, without the involvement of the STS.

Contrarily, entirely different damage is found in patients who suffer from disorders that render it impossible for them to recognize structure based on movement—for example, a human silhouette from points of light. They involve the fusiform gyrus—which participates in facial recognition [29]—and the lingual gyrus, together with the ventrolateral areas of the temporal lobe [20]. The results of those studies combined with functional magnetic resonance imaging (fMRI) examinations—in which it was determined that natural stimuli caused excitation in both the ventral and the dorsal pathways, while "light points" did so only in the ventral pathway—suggest that the ventral pathway is primarily involved in deducing structure on the basis of perceived movement.

9.2.3 *Perception of Implied Motion*

When we see a moving object "paused" on a photograph or a digital screen, we automatically interpret what we are seeing as a sample of a space–time sequence. That we have no problem telling that the objects shown (which are, after all, still) are "flying" or "jumping" indicates that the process of deduction concerning perceived movement, which is known as deduced movement, is an automatic one. The automatic nature of the processes of perceiving deduced movement is evidenced by studies in which it was proven that humans tend to consider photographs taken one after the other over a short period of time as identical [21].

Studies of this phenomenon are conducted by applying different types of stimulus. Geisler [22] demonstrated that humans use traces of movement as indicators that make it easier to detect moving objects. It is likely that the cells of the primary visual cortex are responsible for this function. Moreover, Ross et al. [23] proved that it is possible to perceive movement based solely on an object's form. Both of the above studies utilized Glass patterns—layouts of pairs of points arranged along a predetermined "path", which is de facto perceived to be that of their movement. By utilizing specific point arrangements, it is possible to create impressions of different types of movement, such as rotation or moving closer to the viewer. Krekelberg's [24] experiments proved that observation of Glass patterns that suggest rotation causes activation in the MT and MST areas of the superior optic sulcus.

Similar results were presented by Kourtzi and Kanwisher [25], who conducted studies with the use of natural photographs of moving objects. The cortical areas involved in the perception of movement are also active when perceiving static images that display objects or individuals in motion. Observation of photographs of moving objects causes stronger activations in both the MT and MST than that of the same objects photographed at rest. This effect occurs both in the case of animate objects (e.g. basketball players) and inanimate objects (e.g. sea waves), and is reproducible for diverse categories of objects. The authors present two possible interpretations of this phenomenon.

The results of the above studies suggest that in addition to possessing the capacity to perceive movement, humans also find it easy to spot motion when observing seemingly still objects; more importantly, the processes of perceiving such motion involve the areas associated with perception of movement, including the superior temporal sulcus.

This special capacity of the human visual apparatus to perceive movement is also used in human–computer interactions, and is the subject of a significant number of studies.

9.2.4 The Mirror Mechanism in Action Perception

Humans are social creatures, and their successful functioning depends on how well they cope with social situations. Understanding other people's actions, emotions or motivations is, to a great extent, automatic and intuitive. But is it possible to reveal a general mechanism that is responsible for this process? Some might assert that understanding of other people materializes on the cognitive or declarative levels, as we have insight into our own thoughts. Beyond the plane of conscious judgment, however, there is also a mechanism that operates, as stated above, automatically. Most generally, this could be called a "mirror mechanism" [26]. The architects of the concept claim that understanding of the social world is based largely on the phenomenon of brain "imaging" of others' activities or emotions, which is based on the brain's capacity to link what we ourselves experience with what we observe, and with what is experienced by others.

In the classic approach [27], perception of others' actions is based upon perceiving the individual elements and what is happening to them—for example, when we see a child reaching for a glass of milk, we can see the child, the glass of milk, the movement of the hand, and the movement of the glass; linking these types of information together allows us to understand the perceived action. In recent years, however, an entirely different understanding of the process has gained currency. The first proponents of this new model appeared after Rizzollatti et al. [28] had discovered mirror neurons. When studying rhesus macaques, Rizzollatti discovered that the neurons located in the ventral premotor cortex (in the F5 area) were activated both when the monkeys were performing actions—such as reaching for an object—and when they perceived the same action performed by others. The discovery is rumored to have occurred partially "by accident", when one of the researchers reached for a piece of fruit in a macaque's field of vision, and then noted increased activity on the diagram—similar to that when the monkey itself had been reaching for the fruit. The primary assumption of the model is that perceiving actions activates the same areas as performing the same actions; and, furthermore, that recognizing the actions of others is not based on the activation of visual representations (which, of course, also occurs), but on "experiencing" the actions perceived and activation of the observer's "motor knowledge", as Gallese terms it.

The existence of such a mechanism, and the suggestion that it might contribute to perception, is also applicable to human–computer interaction. This opens new and interesting perspectives for researchers, and offers new opportunities to computer system developers who strive to increase application performance, while simultaneously ensuring the greatest degree of user comfort. The effects associated the mirror system will also become increasingly relevant as immersive reality, in which interactions with others happen via humanoid avatars or computer-controlled avatars, gains in popularity. Although such avatars cannot be described as "new", VR offers a higher degree of realism, and the natural biological effects of human perception are set to emerge all the more strongly.

9.3 Perception of Movement in Human–Computer Interactions

The problems related to the role of motion in human–computer interactions involving a screen were discussed by researchers as early as the end of the twentieth century [29]. They postulated that canny use of animations and the moving elements of interfaces might benefit users' comfort. If changes that occur in an interface are so sudden that the user cannot easily follow where they begin and end, the cause-and-effect relationship of the system will elude him/her. The user will begin to consider how the objects displayed on the screen are related to those that were there previously; it will be unclear to him/her which changes result from his/her actions, and which have not. In the last 20 years, animations have been used in the development of mobile and desktop operating systems to illustrate the changes occurring in their statuses, superseding the sudden shifts in the graphic user interface that were once commonplace. Animations have also been applied to improve user comfort by creating an illusion of greater smoothness in the operation of systems. This was utilized, for instance, in smartphones equipped with the Windows Phone system.

9.3.1 Using Movement to Influence Interaction Flow

Intriguing studies on how moving elements on a screen might affect users' opinion on their interaction with the system were conducted by Harrison et al. [30, 31]. Commencing with the assumption that human perception of the passage of time was subjective, and that perception of a user interface might be subject to certain illusions [32], they proposed employing animations to influence users' perception of systems' performance. The studies focused on the manner of presentation of progress bars, which constitute an integral element of user interfaces [33]. The research demonstrated that the right modification of the progress bar's movement speed could contribute to increasing an interface's usefulness. This carries particular importance for illustrating progress types whose duration might initially prove challenging to determine, such as hard drive repair/defragmentation, or downloading large files over an unstable internet connection. Implementing the right model of progress bar movement alleviates the irritation of slowdown or pauses when progress bars are nearing 100%. Artificially decreasing a progress bar's speed of movement upon the commencement of an operation, and increasing it when it nears completion, was perceived positively by users. Meanwhile, slowdowns and pauses were much better tolerated when they occurred at the beginning of the process. The authors of the studies suggest considering this when designing interactions and place elements that require more time at the beginning of multi-stage processes, such as the installation of software.

9.3.2 Emotion and the Perception of Affective Qualities of Interface

In neurobiological terms, emotions are associated with movement and the perception of movement. It has been proven that the dynamics of movement influences the perception of emotional expressions' intensity [34]; thus, we might reasonably expect in the case of human interaction with computers, that the movement of elements—on a screen, for instance—might have an effect on the user's emotions, too. The potentially crucial role played by motion in human–computer interactions was acknowledged more than 20 years ago [35, 36]. Despite the passage of time, however, studies in the field remain scant. Those interested in designing robots analyze the relationship between movement and emotion and how it can be used in practice, with the highest frequency. Saerbeck and Bartneck [37] tested how the parameters of robots' movements impacted how they were perceived emotionally. They studied the perception of two types of robot: the iCat, which was designed specifically for studying interactions; and the Roomba, a typical household cleaning robot. The movement parameters subjected to analysis were acceleration and curvature, and the relative variables were arousal and emotion valence. The strongest relationship was observed between acceleration and arousal, but it was also discovered that none of the movement parameters analyzed affected the perception of emotions' valence. This means that while designing interaction that involves movement, impressions of different levels of user arousal can be obtained by manipulating the acceleration parameter. The studies discussed here focused on interactions with robots; nevertheless, the results obtained spring from basic principles of emotional psychology, and they also apply to other types of interaction—particularly since the study discovered that the relationship between emotions and the physical parameters of movement was the same for both of the robots. That relationship transpired to be universal.

9.3.3 Modelling Attitudes with Perceived Motion

Perception of movement is a key factor in the field of human–robot interaction. Studies on the influence of perceived motion on interactions with robots have been conducted by Hoffman and Vanunu [38]. They built a robotic speaker that could synchronize with the music it played, allowing the possibility of interaction with it. In experiments testing how such a robot would be received by users, how its movements would influence participants' enjoyment of music and their opinions on it, was studied. It was revealed that appropriate movements of the robot (those which were consistent with the song's rhythm) could increase the attractiveness of the tracks to which the participants listened. This result suggests that human attitudes are modeled not only by observing the behaviors of others—as postulated by Bandura [39] in his Social Learning theory—but also through tailored movements of robots. It seems reasonable that analogous effects would be observed in the case of avatars displayed during interactions involving screens or occurring in VR. Such solutions would enable their designers to influence users' reactions beyond the confines of multimedia entertainment. By executing their designers' intentions, avatar reactions could attempt to influence user behavior by, for instance, shaping positive reactions to events occurring during interactions. The effectiveness of this type of mechanism is founded on the presence of mirror neurons, which allows us to perceive the social world. The process of perceiving the behaviors of other people and robots, as demonstrated in the example above, operates with the involvement of the same brain structures that are responsible for performing the given actions.

9.3.4 Perceived System Performance and User Interface Animations

Appropriate design of interactions containing moving elements can influence not only the subjective comfort of systems' use, but also their perceived performance. Garcia et al. [40] conducted studies of perception of the performance of handwriting recognition and part-of-speech tagging systems. In a series of tests, users were presented either with animations displaying a system's operation, in which words of the text analyzed or labels of parts of speech appeared in animated fashion, or with static images of the same content. It turned out that when the system was presented in an animated way, its effectiveness was rated higher than in the case of static presentations. It is also worth noting that the stimuli used in the study deliberately included an error, but it remained unnoticed - the positive effect of the movement of words appearing on the screen was strong enough to offset the potential negative effect of the error.

9.3.5 Motion and Intelligibility of Computer Systems

The development of ubiquitous computing systems, or those based on artificial intelligence (AI) algorithms, results in elements of their operation being carefully concealed from the user. This can lead to situations in which users find systems' operation surprising, which, in turn, reduces the comfort of their use. In some circumstances—namely, when these surprises result from machine learning solutions, it might be impossible to share such knowledge with the user. In the case of ubiquitous computing systems, whose operation is based on data collected without user input, lack of understanding can be solved by implementing two parameters of interaction: intelligibility and accountability [41]. Systems should communicate how they "perceive" the world, and they should allow their users control to make decisions when necessary. How might this be implemented? Vermulen explored the potential of presentations that use motion to increase the intelligibility and control of ubiquitous computing systems. In pilot studies, he proved that displaying animations that explain a system's operation might effectively contribute to users' understanding of it. Methods of this type can also be used to visualize the results of planned actions.

9.3.6 *Effectiveness of Animation in Research Results Dissemination*

A primary function of science is the dissemination of findings. In the case of researchers, this usually manifests in the form of scientific papers or presentations delivered during conferences. Another key area, however, lies in the dissemination of the knowledge obtained through studies among members of the groups to which it is of concern. This is particularly relevant in the field of health, for example. Some researchers have proposed the creation of animated materials to facilitate the transfer of knowledge to concerned extra-scientific communities. With the assistance of participatory design methodology, Voughn [42] and his colleagues have created dissemination materials for projects dedicated to the prevention of youth violence.

9.3.7 *Perception of Motion of Avatars*

One area in which the perception of movement naturally plays an important role is in the use of avatars—graphic representations of users. On internet forums, social media, and in other communities, avatars are typically presented in the form of two-dimensional images; in games and in virtual worlds, however, they are more commonly three-dimensional. Avatars allow users to create digital representations of themselves that are visible to their friends. Avatars are also used in video games as representations of protagonists that are controlled by players. From the perspective of this chapter, three-dimensional animated avatars—which, aside from their distinctive features of appearance, also have a specific way of moving—are of most relevance. Such avatars might be present in VR systems; they can be displayed on computer screens, or constitute a part of the "real" world—such as in the case of the Millie virtual sales assistant, which utilizes a camera system, and is displayed on screens in retail stores.

A pressing problem concerning the movement of avatars lies in synchronizing the movements that users makes in virtual worlds and the motion that they perceive to be that of the virtual avatars. Choi et al. [43] suggest that this of great importance. They compared the impressions of study participants with the use of different methods of motion imaging, which differed with respect to how well they synchronized with the users' movements. Better synchronization reduced feelings of cybersickness [44], increased the participants' sense of presence, and the impression that they were a part of the virtual world. It also served to enhance the participants' impression of having a virtual body (body ownership). This is likely linked to the participants recognizing the movements of their own bodies based on their kinematic specifics [45]. The ability to correctly recognize one's own body, both when watching point-light displays and the movement of virtual characters, is well established. Interestingly, in the case of avatars, the same effect applied when they had been deprived of all features related to the true appearance of their users, and even when the avatar presented was of a different sex than its user. This demonstrates that the way an avatar moves might enable the identification of individuals. Since it is possible to recognize oneself in an avatar's motions, perhaps it is also possible to recognize others. With this in mind, it is easy to imagine the use of a "kinematic signature" of sorts in virtual worlds, containing information on the specifics of a user's movements which, when applied to a virtual character, would enable the creation of an avatar that moves in a way characteristic for and unique to a specific user.

References

1. Bassili, J.N.: Facial motion in the perception of faces and of emotional expression. J. Exp. Psychol. Hum. Percept. Perform. **4**, 373–379 (1978)
2. Bassili, J.N.: Emotion recognition: the role of facial movement and the relative importance of upper and lower areas of the face. J. Pers. Soc. Psychol. **37**, 2049–2058 (1979)
3. Van Boxtel, J.J.A., Lu, H.: A biological motion toolbox for reading, displaying, and manipulating motion capture data in research settings. J. Vis. **13** (2013). https://doi.org/10.1167/13.12.7
4. Ma, Y., Paterson, H.M., Pollick, F.E.: A motion capture library for the study of identity, gender, and emotion perception from biological motion (2006). https://doi.org/10.3758/bf03192758
5. Bachynskyi, M., Oulasvirta, A., Palmas, G., Weinkauf, T.: Is motion capture-based biomechanical simulation valid for HCI studies? study and implications. In: Proceedings of the SIGCHI Conference on Human Factors in Computing Systems, pp. 3215–3224. Association for Computing Machinery, New York, NY, USA (2014)
6. Flindall, J.W., Gonzalez, C.L.R.: Revisiting Ungerleider and Mishkin: two cortical visual systems (2017).https://doi.org/10.4135/9781529715064.n5.
7. Goodale, M.A., Milner, A.D.: Separate visual pathways for perception and action. Trends Neurosci. **15**, 20–25 (1992)
8. Giese, M.A., Poggio, T.: Neural mechanisms for the recognition of biological movements. Nat. Rev. Neurosci. **4**, 179–192 (2003)
9. Bobick, A.F.: Movement, activity and action: the role of knowledge in the perception of motion. Philos. Trans. R. Soc. Lond. B Biol. Sci. **352**, 1257–1265 (1997)
10. Todd, J.T.: Perception of gait. J. Exp. Psychol. Hum. Percept. Perform. **9**, 31–42 (1983)
11. Beintema, J.A., Lappe, M.: Perception of biological motion without local image motion. Proc. Natl. Acad. Sci. U. S. A. **99**, 5661–5663 (2002)
12. Downing, P.E., Jiang, Y., Shuman, M., Kanwisher, N.: A cortical area selective for visual processing of the human body. Science **293**, 2470–2473 (2001)
13. Hodzic, A., Kaas, A., Muckli, L., Stirn, A., Singer, W.: Distinct cortical networks for the detection and identification of human body. Neuroimage **45**, 1264–1271 (2009)
14. Oram, M.W., Perrett, D.I.: Integration of form and motion in the anterior superior temporal polysensory area (STPa) of the macaque monkey. J. Neurophysiol. **76**, 109–129 (1996)

15. Howard, R.J., Brammer, M., Wright, I., Woodruff, P.W., Bullmore, E.T., Zeki, S.: A direct demonstration of functional specialization within motion-related visual and auditory cortex of the human brain. Curr. Biol. **6**, 1015–1019 (1996)
16. Bonda, E., Petrides, M., Ostry, D., Evans, A.: Specific involvement of human parietal systems and the amygdala in the perception of biological motion. J. Neurosci. **16**, 3737–3744 (1996)
17. Decety, J., Grèzes, J.: Neural mechanisms subserving the perception of human actions. Trends Cogn. Sci. **3**, 172–178 (1999)
18. Vaina, L.M., Lemay, M., Bienfang, D.C., Choi, A.Y., Nakayama, K.: Intact "biological motion" and "structure from motion" perception in a patient with impaired motion mechanisms: a case study. Vis. Neurosci. **5**, 353–369 (1990)
19. McLeod, P.: Preserved and impaired detection of structure from motion by a "motion-blind" patient (1996). https://doi.org/10.1080/135062896395634
20. Cowey, A., Vaina, L.M.: Blindness to form from motion despite intact static form perception and motion detection. Neuropsychologia **38**, 566–578 (2000)
21. Freyd, J.J.: The mental representation of movement when static stimuli are viewed. Percept. Psychophys. **33**, 575–581 (1983)
22. Geisler, W.S.: Motion streaks provide a spatial code for motion direction. Nature **400**, 65–69 (1999)
23. Ross, J., Badcock, D.R., Hayes, A.: Coherent global motion in the absence of coherent velocity signals. Curr. Biol. **10**, 679–682 (2000)
24. Krekelberg, B., Dannenberg, S., Hoffmann, K.-P., Bremmer, F., Ross, J.: Neural correlates of implied motion. Nature **424**, 674–677 (2003)
25. Kourtzi, Z., Kanwisher, N.: Activation in human MT/MST by static images with implied motion. J. Cogn. Neurosci. **12**, 48–55 (2000)
26. Gallese, V., Keysers, C., Rizzolatti, G.: A unifying view of the basis of social cognition. Trends Cogn. Sci. **8**, 396–403 (2004)
27. Fodor, J.A.: The Modularity of Mind. MIT Press (1983)
28. Rizzolatti, G., Fadiga, L., Gallese, V., Fogassi, L.: Premotor cortex and the recognition of motor actions. Brain Res. Cogn. Brain Res. **3**, 131–141 (1996)
29. Chang, B.-W., Ungar, D.: Animation: from cartoons to the user interface. In: Proceedings of the 6th Annual ACM Symposium on User Interface Software and Technology, pp. 45–55. Association for Computing Machinery, New York, NY, USA (1993)
30. Harrison, C., Yeo, Z., Hudson, S.E.: Faster progress bars: manipulating perceived duration with visual augmentations. In: Proceedings of the SIGCHI Conference on Human Factors in Computing Systems, pp. 1545–1548. Association for Computing Machinery, New York, NY, USA (2010)
31. Harrison, C., Amento, B., Kuznetsov, S., Bell, R.: Rethinking the progress bar. In: Proceedings of the 20th Annual ACM Symposium on User Interface Software and Technology, pp. 115–118. Association for Computing Machinery, New York, NY, USA (2007)
32. Tognazzini, B.: Principles, techniques, and ethics of stage magic and their application to human interface design. In: Proceedings of the INTERACT '93 and CHI '93 Conference on Human Factors in Computing Systems, pp. 355–362. Association for Computing Machinery, New York, NY, USA (1993)
33. Myers, B.A.: The importance of percent-done progress indicators for computer-human interfaces. SIGCHI Bull. **16**, 11–17 (1985)
34. Biele, C., Grabowska, A.: Sex differences in perception of emotion intensity in dynamic and static facial expressions. Exp. Brain Res. **171**, 1–6 (2006)
35. Vaughan, L.C.: Understanding movement. In: Proceedings of the ACM SIGCHI Conference on Human Factors in Computing Systems, pp. 548–549 (1997)
36. Bacigalupi, M.: Designing movement in interactive multimedia: making it meaningful. ACM (2001)
37. Saerbeck, M., Bartneck, C.: Perception of affect elicited by robot motion. In: 2010 5th ACM/IEEE International Conference on Human-Robot Interaction (HRI), pp. 53–60 (2010)

38. Hoffman, G., Vanunu, K.: Effects of robotic companionship on music enjoyment and agent perception. In: 2013 8th ACM/IEEE International Conference on Human-Robot Interaction (HRI), pp. 317–324 (2013)
39. Bandura, A.: Social Learning Theory. Prentice Hall (1977)
40. García García, P., Costanza, E., Verame, J., Nowacka, D., Ramchurn, S.D.: Seeing (movement) is believing: the effect of motion on perception of automatic systems performance. Hum. Comput. Interact. **36**, 1–51 (2021)
41. Bellotti, V., Edwards, K.: Intelligibility and accountability: human considerations in context-aware systems. Hum. Comput. Interact. **16**, 193–212 (2001)
42. Vaughn, N.A., Jacoby, S.F., Williams, T., Guerra, T., Thomas, N.A., Richmond, T.S.: Digital animation as a method to disseminate research findings to the community using a community-based participatory approach. Am. J. Community Psychol. **51**, 30–42 (2013)
43. Choi, C., Jun, J., Heo, J., Kim, K. (kenny): Effects of virtual-avatar motion-synchrony levels on full-body interaction. In: Proceedings of the 34th ACM/SIGAPP Symposium on Applied Computing, pp. 701–708. Association for Computing Machinery, New York, NY, USA (2019)
44. Kemeny, A., Chardonnet, J.-R., Colombet, F.: Getting Rid of Cybersickness: In Virtual Reality, Augmented Reality, and Simulators. Springer Nature (2020)
45. Thaler, A., Wellerdiek, A.C., Leyrer, M., Volkova-Volkmar, E., Troje, N.F., Mohler, B.J.: The role of avatar fidelity and sex on self-motion recognition. In: Proceedings of the 15th ACM Symposium on Applied Perception, pp. 1–9. Association for Computing Machinery, New York, NY, USA (2018)

Chapter 10
Movement, Cognition and Learning

10.1 Introduction

Since the 1970s, the universalization of computers has increased the amount of time users spend in sitting positions. From an evolutionary perspective, this is highly unnatural, and may lead to a number of so-called "lifestyle diseases". This is proven by anthropological studies of contemporary hunter-gatherer cultures, whose lifestyles resembles those of the majority of humans for the 150,000 years preceding the emergence of agriculture. In the tribes of this type, such as the !Kung, short periods of rest are interspersed with longer periods of searching for sustenance, due to permanent difficulties in obtaining food. Anthropologists estimate that the average daily distance covered by members of contemporary hunter-gatherer tribes ranges between 10 (for women) and 20 km (for men) [1]. There are reasons to believe that the need for physical activity is encoded in our genes [2]. As the primary reason for which humans once undertook activity was to obtain food—a task that also involved mental processes—physical activity remains closely associated with cognitive capabilities. Some authors believe that thinking is movement internalized [3]. The impact of physical activity, therefore, is not limited to physical health; much evidence asserts that it also has a positive influence on mental functioning [4, 5].

Presently, it is difficult to imagine such a drastic change in the lives of modern societies that would compel humans to walk 10–20 km a day. It has transpired, however, that much less activity may be required to instigate positive effects in mental functioning. Studies demonstrate that exercising for 30 min leads to a noticeable improvement in memory and motor coordination [6]. Thirty minutes of aerobic activity has been revealed to be a factor that improves cognitive functions in those suffering from depression. It has also been shown that short, three-minute sprints improve memorization [7].

With the development of technologies such as smartphones and virtual reality (VR), a gradual shift can be observed in the paradigm of computer use from desktop to mobile. There is an opportunity, therefore, for computers, as one the key causes of

C. Biele, *Human Movements in Human-Computer Interaction (HCI)*,
Studies in Computational Intelligence 996,
https://doi.org/10.1007/978-3-030-90004-5_10

sedentary lifestyles (and the health conditions resulting from them), to serve as the cure for the spreading epidemic of idleness. As the subject is also highly intriguing from a commercial perspective, manufacturers of video game consoles have designed solutions to engage gamers in physical activity. Examples include Nintendo with its Wii console, and Sony with its Playstation Move controllers and the EyeToy camera. There are myriad solutions for different forms of physical activity being designed for immersive VR systems, including those based strictly on software, as well on specialized hardware, such as the Icarus or VirZoom systems. With this in mind, it is worth considering precisely how effective computer systems might be in motivating individuals to become physically active, and consequently, in improving their psychological functioning, as well as what conditions exist for such motivation to be effective. The following sections, however, will focus on the relationship between movement and brain functioning.

10.2 Influence of Movement on Cognitive Function

Studies on the influence of physical activity on cognitive function date back to the nineteenth century, such as those by Galton, which were conducted on excitation and its influence on the performance of various tasks. In the mid-twentieth century, the reticular activation system (RAS) [23]—which could potentially be responsible for the aforementioned relationship—was discovered. The RAS is a network of neurons located in the brain stem, which is connected to the cerebral cortex. The network is responsible for regulating states of wakefulness and sleep, as well as for ensuring the appropriate level of excitation during waking hours. Activity of the RAS decreases with age, which is deemed to be the reason for the differences observed in a number of experiments focused on reaction times: the reactions of sixty-year-olds are at least 20–50% slower than those of younger individuals [24, 25]. Knowing that RAS activity is associated with an individual's momentary activity level, it can be concluded that the sitting position is the least conducive to solving mental tasks. According to the Yerkes-Dodson law [26], a relationship in the shape of an inverted "U" should be visible—i.e. both too small and too large an excitation of the RAS would have a negative impact on the performance of mental tasks. The potential existence of this type of mechanism is corroborated by studies involving individuals with pacemakers: external increases in the frequency of heart contractions led to better performance of mental tasks [27].

The first studies to test the hypothesis of the relationship between body position (sitting, standing, prone) and reaction speed were conducted by Woods and her colleagues [28]. The studies involved both younger (aged 18–28) and older (aged 60–70) subjects. In the case of the younger participants, no relationship was observed between body position while performing tasks and the reaction times displayed during those tasks; while differences were observed among the older participants. The fastest reactions were recorded from a standing position. To verify the influence of fitness on the results of the study, the participants were subsequently divided into two groups. The division was based on the results obtained in a stationary bicycle stress test. During the tests, fitter individuals demonstrated shorter reaction times. Following the division into subgroups, the largest influence of body position on reaction times was recorded in the cases of older, unfit individuals. This demonstrates that age and poor fitness are additive factors that affect the performance of mental tasks. The largest negative impact of the sitting position on cognitive functioning was recorded in the case of older participants who, by using a computer in a traditional manner, further lowered their fitness levels, and simultaneously, reducing their RAS activation. It seems, therefore, that it is particularly important that technological solutions be developed which are targeted specifically at the elderly. Woods also demonstrated that increased excitation resulting from physical exercise may improve cognitive functioning; although this relationship depends on numerous factors, such as age, task difficulty, and physical condition.

In the case of schoolchildren, implementing a short interval exercise regimen during classes—in the form of intense, 20 s stages interspersed with 10 s breaks—had a positive influence on their attention span [29]. Similarly, the introduction of short, 10 min breaks during classes, in which students performed aerobic exercises, had a positive effect on their engagement in the tasks performed during classes [30]. This effect was especially noticeable in the group of students who usually displayed low engagement in classwork. Physical activity (and movement itself) is also a factor conducive to teaching physical concepts related to mechanics [31]. Students who had the opportunity to participate physically in mechanics experiments memorized the concepts presented much better than those who did not. This is consistent with the theory of embodied cognition—according to which mental processes do not involve only the mind, but the whole body [32].

Even a short period of exercise can influence how the human brain works. There is extensive evidence proving that even a single training session may have a positive influence on cognitive function and enhance working memory, for instance [33]. These effects last up to one hour following the exercise. Similar results were obtained for other indicators, such as cognitive flexibility [34].

The positive influence of physical activity on cognitive functions is also evidenced indirectly by studies of school achievements conducted in Italy [35]. These studies verified possible relationships between the results of physical fitness tests and academic achievement. The factors accounted for when evaluating fitness included muscle strength of the lower and upper limbs, running results, and agility (jumping, among other things). The running results turned out to be related to grades in English, Italian, mathematics, and music; while jumping results were found to be related to grades in English, mathematics, and technology. It is reasonable to conclude from the above that factors that increase physical activity and improve performance in sports may also contribute to superior academic achievements.

All of the above results demonstrate clearly that if societies aim to foster the mental development of young people and to ensure the effective functioning of the elderly, they must seek to increase the physical activity of those groups. Could technology prove helpful in achieving this objective?

10.3 Brain Mechanisms Underlying the Connection Between Physical Activity and Cognition

Physical activity affects mental function via a number of parallel mechanisms, such as: changes in electrical activity and blood flow in the attention and memory systems; function changes on the level of neurotransmitters; and changes on the level of neurons (e.g. neurogenesis) [4].

As for the functioning of the attention and memory systems, the electrophysiological indicator of the effectiveness of information processing is the P300 event-related brain potential (ERP) component—a reaction that occurs approximately 300–800 ms after receiving a stimulus related to its attention-related processing. The latency of P300 potential is related to the speed of a given stimulus' processing, while the amplitude points to the resources necessary to execute the process [8]. Following physical exercise, this amplitude increases and the latency of P300 potential is reduced [9]. Fitter people demonstrate higher P300 amplitudes and shorter latencies [10]. The results of the studies cited indicate clearly that physical activity contributes to the effective functioning of cerebral attention-related systems. Similarly positive impacts have been observed in relation to the brain areas responsible for memory processes.

Although constituting a mere 2% of its mass, the brain consumes 20% of a human body's energy [11]. As a result of physical exercise, the blood flow and the transport efficiency of oxygen and glucose to the brain increase, which improves its functioning [12]. By utilizing functional magnetic resonance imaging (fMRI), Pereira et al. [13] demonstrated that physical exercise increases the blood flow in the hippocampus—a key area of the brain associated with memory function—and, in consequence, affects the results of memory tests. After a training regimen, the study participants' results improved, arguably due to improved blood circulation leading to more effective transport of nutrients and oxygen to the brain.

The two best understood chemicals responsible for the effects of physical activity on cognitive functioning are brain derived neurotrophic factor (BDNF) and insulin-like growth factor 1 (IGF-1). Studies show that physical activity leads to increased levels of BDNF in the hippocampus. After experiments were conducted on mice, it was discovered that physical exercise improved the animals' performance in the Morris water maze; while blocking the action of BDNF with drugs negated the positive impact of exercise on their performance [14].

IGF-1 is a factor that mediates the action of growth hormones and the remodeling of body tissues. This includes the brain, where IGF-1 is produced. The hormone is also capable of passing through the blood–brain barrier, which allows it to reach the brain from the circulatory system, while physical exercise increases IGF-1 uptake in, among other areas, the hippocampus. In humans, IGF-1 levels correlate positively with cognitive function [15].

Another reason that substances such as BDNF and IGF-1 have positive effects on cognitive function is that they are factors that support the creation of new neurons, in addition to contributing to the improvement of blood circulation. Formation of new neurons in the hippocampus area, which is one of the primary brain structures responsible for memory, increases proportionately to the amount of physical activity performed [16].

Physical activity also affects a number of processes that occur in the human brain through neurotransmitters such as serotonin, dopamine, and norepinephrine. Physical exercise increases the release of those substances, which in turn participate in the processes of adjusting an individual's mood and motivation [17]. It is known, for example, that physical exercise enhances positive emotions [18], improves the mood [19], and reduces stress levels [20]. Furthermore, good mood and lack of stress are key factors for effective learning—learning is complemented by positive emotions [21]. The prefrontal cortex, which is responsible for higher cognitive functions, ceases to work optimally in the presence of stress hormones, such as cortisol or norepinephrine [22].

As demonstrated above, much knowledge has already been collected on the cerebral mechanisms that link physical activity with cognitive function. Although some of the aforementioned studies concerning brain biochemistry were conducted on animals, there are also a host of studies indicating that the same relationship can be observed in humans.

10.4 Using Computers to Promote Physical Activity

The examples mentioned in the previous sections present only a fragment of the knowledge on the impact of physical activity on cognitive functioning. In light of this knowledge, the data regarding global levels of physical activity is alarming. Meta-analyses from 2020, which compile data from nearly 300 studies conducted on more than 1.5 million students around the world, show that more than 80% of the students are insufficiently active [36]. As children develop, physical activity is replaced by sitting [37], a position in which teenagers use computers or smartphones. Can the development of new ways of interaction or inventive use of existing ones contribute to reversing this negative trend; and can technology become an effective tool in promoting physical activity and, consequently, good health, and cognitive performance? This is a crucial question, since current approaches aimed at increasing the levels of physical activity among youths have proven ineffective. This is evidenced by a meta-analysis of 17 interventions (educational and social) which were aimed at either increasing physical activity only or, more broadly, at other pro-health attitudes. All of the interventions had failed to increase the average duration of medium-intensity physical activity [38]. The meta-analysis did not include interventions that involved technology; similar reviews of studies in which technological means were used to increase physical activity among youths, nevertheless, confirmed the low effectiveness of such interventions [39], and the results obtained were ambiguous. It is noteworthy, however, that most of the interventions analyzed, which had been implemented digitally, were presented in the form of websites. Other interventions utilized social media campaigns or text messages, but only a handful were presented in the form of games. It is unfortunate that only a single intervention analyzed in the review concerned physical activity—one which involved interactive races on a stationary bicycle—yet no benefits were observed in utilizing the video game compared to the reference condition, which involved riding a stationary bicycle to the accompaniment of music [40].

10.4.1 Physical Activity in Active Video Games

Games that utilize physical activity, so-called 'exergames' or active video games (AVG), are a potentially effective method of promoting physical activity. Games of this type incite similar physiological benefits as normal physical exercise [41]—they activate attitudinal changes, such as increasing an individual's motivation to undertake physical activity. Energy expenditure in AVGs is comparable to that in traditional physical training [42]. Researchers, however, have noted a problem with the periodic nature of physical exertion: periods of activity (playing the game) are frequently interspersed with breaks, which result from games' loading times [43]. The AVG phenomenon can be utilized effectively for commercial purposes by the manufacturers of video game consoles, for example, in developing sports games. Although used primarily for entertainment and for social purposes [44], it has been proven that applications such as Pokemon GO increase the level of users' physical activity [45]. Interactive games also may increase the motivation for the excersise, they also allow for the remote competition, which also can have positive effect on the motivational aspects of the use of such applicationa (like Zwift).

10.4.2 Engaging Whole Body Movement

Another problem of exergames is that players frequently attempt to minimize their physical exertion, and focus solely on succeeding in the games. A similar phenomenon has been observed regarding social media-based technological solutions—the goal of which was to encourage user activity by sharing the number of steps taken each day. Users rapidly devised special electrical cradles, which could be installed on fitness bands and set to simulate a specific number of steps by being moved in a specific manner. In the case of solutions like the Nintendo Wii, intense physical activity is not required to succeed in games such as tennis. An experienced user may win a game by moving only his/her wrist, while remaining comfortably on a sofa. This demonstrates that external motivation in the form of, for instance, gamification rewards, or even peer recognition on social media, provides insufficient motivation to increase physical activity. Humans' natural tendency to minimize their exertion will lead them to look for ways of doing so whenever possible.

10.4.3 Hardware Solutions for Exergames

In their efforts to create effective games that promote physical activity, exergames developers have frequently turned to solutions based on specially designed hardware elements. This has happened in the case of games designed for children, such as *Bug-Smasher* [46] or *Flash Pole* [47]. One undeniable strength of such systems is the fact that they are attractive to, and engaging for, their users. Due to their physical form, however, they also come with numerous shortcomings: they are unportable, costly, and can usually be used with only a single game—or, in the best case scenario, with several games that, due to the limitations of the hardware, must be highly similar to each another. Solutions in the form of video game consoles with controllers that register movement or attachments, such as the Playstation EyeToy are more universal. The most universal of all solutions, however, which allows users to experience relatively trouble-free tracking of movements of the whole body, is VR. Hybrid solutions also gained popularity especially in 2020 during the worldwide pandemic—Zwift mentioned extended their support also for running, aside from biking, and this application is very popular.

10.4.4 Virtual Reality and Physical Activity Promotion

Virtual reality is a promising area for the development of methods that promote physical activity, and therefore effective cognitive function. The technology could serve to increase motivation, while simultaneously alleviating users' natural tendency to avoid the exertion associated with exergames. A game experienced in VR may contain elements that force users to move their whole bodies. The technology's potential had been tested prior to the release of VR goggles that offer complete immersion, such as the Oculus Quest or the HTC Vive. One such solution is *Astrojumper*, developed by Finkelstein et al. [48]—an arcade game set in space, in which the player must avoid planets rushing at him/her by jumping, ducking, and dodging. The game itself takes place in a room with a stereoscopic image projected onto its walls, while the user's movements are tracked with the assistance of the sensors attached to his/her body. Users' opinions suggested that the game had motivated them to embark on physical exercise. One factor that contributes to exergames' effectiveness as a tool for promoting physical activity is the technology's ability to naturally draw the user's attention away from the physical exertion. In the case of exergames using rowing machines, for example, this factor proved effective. The exercise involved in the game also evoked more positive emotions in its users [49]. It is expected that VR will prove even more effective in this respect. Reports from the initial reactions to immersive VR environments clearly show that users view this experience in terms of "living it" or "being transported to another world", and this reaches beyond the realm of mere perception [50]. Virtual environments are effective in evoking emotions in their users, as demonstrated by Pan et al. [51]. This significantly increases the games' attractiveness, and, as the Sinclair's dual-flow model demonstrates, attractiveness affects the effectiveness of exercise, measured as intensity of the training [52].

10.4.5 Psychological Effects for VR Exercise

Much is known on the influence of physical activity on cognitive function; on how VR can be used to motivate people to exercise, and how effective such training sessions are when compared with physical activity in the real world. Nevertheless, a strong scientific link between these aspects remains absent. Reports are usually limited to highly specific domains: studies of the elderly with cognitive deficits; those involving rehabilitation; or those focused on attenuating the symptoms of depression or depressive disorders. Even in these fields, however, the reports are scarce. A systematic review of the literature from 2020 pertaining to the effects of training sessions in VR (psychological, physiological, and the effectiveness of rehabilitation) returned only four such papers [53]. Additionally, a wide variety of training session types can be implemented in VR, which are not necessarily based on the movement capabilities offered by the technology. One example is that of the aforementioned studies pertaining to depression and anxiety. A typical method applied in the field is virtual reality exposure therapy (VRET). Studies on the influence of movement or physical activity are scant.

One experiment which tested the potential psychological benefits of VR training was included in the studies conducted by Plante et al. [54]. They compared users' reactions to four types of activity: a traditional stroll, a stroll on a treadmill, a stroll on a treadmill in VR, and a virtual stroll in VR (in a sitting position). The stroll on a treadmill combined with VR increased the participants' level of energy the most, contributing to reduced tension and weariness. Similar studies on the effects of cycling on the mood [55] failed to demonstrate any effect of VR on the participants' mood; compared to sitting down, however, exercise did have an effect. The effects of VR were also observed in the case of cognitive functions [56], such as cognitive flexibility and selective attention, which were measured using the Wisconsin Card Sorting Test [57] and the Stroop Test [58]. The Wisconsin Card Sorting Test involves assigning cards according to ever-changing rules, and is used to test cognitive flexibility. The Stroop test, on the other hand, assesses a subject's ability to inhibit cognitive interference—in other words, to react only to certain aspects of stimuli, and to ignore all other aspects. In its most popular implementation, the test involves reacting to words appearing on a screen in different colors (e.g. the word "red" is written in the color blue). Participants are expected to react to the color in which the word is written, while ignoring its meaning. The results of the abovementioned experiments suggest that the VR technology used for enriching physical activity might have a positive influence on both cognitive flexibility and selective attention. Studies conducted on students with the use of immersive VR combined with an exercise bike (VRZoom) show that experiencing VR during physical training has a positive influence on users' self-perception of both their capabilities and subjective pleasure [59].

While analyzing the results of the aforementioned studies, it should be remembered, however, that they were not all conducted with the use of systems that ensured total immersion. Furthermore, the type of movement studied had a highly limited and specific character (such as that on a stationary bicycle or a treadmill), and failed to address interactions with the environment. These studies proved unable to tap the full potential of VR in encouraging their participants to move.

10.5 Virtual Reality in Training and Rehabilitation

10.5.1 Movement in Virtual Reality in Education

With its possibilities for interaction through movement, VR seems ideally suited to teaching in areas in which movement is an important element, such as that of training surgeons, for instance. The meta-analyses of Alaker et al. and Jin et al., both encompassing more than 30 papers concerning the effectiveness of VR training sessions in laparoscopic surgery, fail to offer conclusive answers on the effectiveness of VR [60, 61].

The utility of VR as a tool for practicing manual skills has also been tested in the training of welders [62]. The results of those studies show that individuals who participated in VR training sessions obtained results that were similar to, and in certain cases, superior to, those of participants in traditional training courses.

10.5.2 Movement in Virtual Reality in Rehabilitation

Studies have suggested that VR might prove helpful in the process of regaining upper limb control lost as a result of a stroke. Stroke patients whose rehabilitation sessions involved VR regained more arm control than those who underwent traditional rehabilitation sessions [63].

VR is also useful in alleviating the phantom limb phenomenon, experienced by amputees. Sensors mounted on the limb stump read its movements and translate it into movements of a virtual limb displayed in VR. Here, VR is an evolution of the previously-used analogue method, which incorporated mirrors to create an optical illusion, resulting in the healthy remaining limb being visible in the place of the amputated one [64, 65]. Modern VR-based solutions utilize head-mounted displays, electromyography sensors, or Kinect sensors [66].

References

1. Cordain, L., Gotshall, R.W., Eaton, S.B., Eaton, S.B., 3rd.: Physical activity, energy expenditure and fitness: an evolutionary perspective. Int. J. Sports Med. **19**, 328–335 (1998)
2. Booth, F.W., Chakravarthy, M.V., Gordon, S.E., Spangenburg, E.E.: Waging war on physical inactivity: using modern molecular ammunition against an ancient enemy. J. Appl. Physiol. **93**, 3–30 (2002)
3. Llinas, R.R.: I of the Vortex: From Neurons to Self. MIT Press (2002).
4. Ratey, J.J., Loehr, J.E.: The positive impact of physical activity on cognition during adulthood: a review of underlying mechanisms, evidence and recommendations. Rev. Neurosci. **22**, 171–185 (2011)
5. Kramer, A.F., Erickson, K.I.: Capitalizing on cortical plasticity: influence of physical activity on cognition and brain function. Trends Cogn. Sci. **11**, 342–348 (2007)
6. McDonnell, M.N., Buckley, J.D., Opie, G.M., Ridding, M.C., Semmler, J.G.: A single bout of aerobic exercise promotes motor cortical neuroplasticity. J. Appl. Physiol. **114**, 1174–1182 (2013)
7. Winter, B., Breitenstein, C., Mooren, F.C., Voelker, K., Fobker, M., Lechtermann, A., Krueger, K., Fromme, A., Korsukewitz, C., Floel, A., Knecht, S.: High impact running improves learning. Neurobiol. Learn. Mem. **87**, 597–609 (2007)
8. Kramer, A.F., Strayer, D.L.: Assessing the development of automatic processing: an application of dual-task and event-related brain potential methodologies. Biol. Psychol. **26**, 231–267 (1988)
9. Hillman, C.H., Snook, E.M., Jerome, G.J.: Acute cardiovascular exercise and executive control function. Int. J. Psychophysiol. **48**, 307–314 (2003)
10. Kamijo, K., Takeda, Y.: Regular physical activity improves executive function during task switching in young adults. Int. J. Psychophysiol. **75**, 304–311 (2010)
11. Raichle, M.E., Gusnard, D.A.: Appraising the brain's energy budget. Proc. Natl. Acad. Sci. U. S. A. **99**, 10237–10239 (2002)
12. Delp, M.D., Armstrong, R.B., Godfrey, D.A., Harold Laughlin, M., David Ross, C., Keith Wilkerson, M.: Exercise increases blood flow to locomotor, vestibular, cardiorespiratory and visual regions of the brain in miniature swine (2001). https://doi.org/10.1111/j.1469-7793.2001.t01-1-00849.x
13. Pereira, A.C., Huddleston, D.E., Brickman, A.M., Sosunov, A.A., Hen, R., McKhann, G.M., Sloan, R., Gage, F.H., Brown, T.R., Small, S.A.: An in vivo correlate of exercise-induced neurogenesis in the adult dentate gyrus. Proc. Natl. Acad. Sci. U. S. A. **104**, 5638–5643 (2007)
14. Gomez-Pinilla, F., Vaynman, S., Ying, Z.: Brain-derived neurotrophic factor functions as a metabotrophin to mediate the effects of exercise on cognition: Exercise and brain metabolism. Eur. J. Neurosci. **28**, 2278–2287 (2008)
15. Arwert, L.I., Deijen, J.B., Drent, M.L.: The relation between insulin-like growth factor I levels and cognition in healthy elderly: a meta-analysis. Growth Horm. IGF Res. **15**, 416–422 (2005)
16. Van Praag, H., Shubert, T., Zhao, C.: Exercise enhances learning and hippocampal neurogenesis in aged mice. J. Neurosci. (2005)
17. Basso, J.C., Suzuki, W.A.: The Effects of acute exercise on mood, cognition, neurophysiology, and neurochemical pathways: a review. Brain Plast. **2**, 127–152 (2017)
18. Reed, J., Ones, D.S.: The effect of acute aerobic exercise on positive activated affect: a meta-analysis. Psychol. Sport Exerc. **7**, 477–514 (2006)
19. Maroulakis, E., Zervas, Y.: Effects of aerobic exercise on mood of adult women. Percept. Mot. Skills. **76**, 795–801 (1993)
20. Ebbesen, B.L., Prkachin, K.M., Mills, D.E., Green, H.J.: Effects of acute exercise on cardiovascular reactivity. J. Behav. Med. **15**, 489–507 (1992)
21. Diamond, A.: The evidence base for improving school outcomes by addressing the whole child and by addressing skills and attitudes not just content. Early Educ. Dev. **21**, 780–793 (2010)

22. Birnbaum, S., Gobeske, K.T., Auerbach, J., Taylor, J.R., Arnsten, A.F.T.: A role for norepinephrine in stress-induced cognitive deficits: α-1-adrenoceptor mediation in the prefrontal cortex. Biol. Psychiatry. **46**, 1266–1274 (1999)
23. Lindsley, D.B., Schreiner, L.H., Knowles, W.B., Magoun, H.W.: Behavioral and EEG changes following chronic brain stem lesions in the cat. Electroencephalogr. Clin. Neurophysiol. **2**, 483–498 (1950)
24. Birren, J.E., Fisher, L.M.: Aging and speed of behavior: possible consequences for psychological functioning. Annu. Rev. Psychol. **46**, 329–353 (1995)
25. Salthouse, T.A.: Aging and measures of processing speed. Biol. Psychol. **54**, 35–54 (2000)
26. Yerkes, R.M., Dodson, J.D., Others: the relation of strength of stimulus to rapidity of habit-formation. Punishm. Issues Exp. 27–41 (1908)
27. Lagergren, K.: Effect of exogenous changes in heart rate upon mental performance in patients treated with artificial pacemakers for complete heart block. Br. Heart J. **36**, 1126–1132 (1974)
28. Woods, A.M., Vercruyssen, M., Birren, J.E.: Age differences in the effect of physical activity and postural changes on information processing speed. In: Proceedings of the Human Factors and Ergonomics Society 37th Annual Meeting, pp. 177–181. Human Factors and Ergonomics Society (1983)
29. Ma, J.K., Le Mare, L., Gurd, B.J.: Four minutes of in-class high-intensity interval activity improves selective attention in 9- to 11-year olds. Appl. Physiol. Nutr. Metab. **40**, 238–244 (2015)
30. Mahar, M.T., Murphy, S.K., Rowe, D.A., Golden, J., Shields, A.T., Raedeke, T.D.: Effects of a classroom-based program on physical activity and on-task behavior. Med. Sci. Sports Exerc. **38**, 2086–2094 (2006)
31. Kontra, C., Lyons, D.J., Fischer, S.M., Beilock, S.L.: Physical experience enhances science learning. Psychol. Sci. **26**, 737–749 (2015)
32. Foglia, L., Wilson, R.A.: Embodied cognition (2013). https://doi.org/10.1002/wcs.1226
33. Lambourne, K., Tomporowski, P.: The effect of exercise-induced arousal on cognitive task performance: a meta-regression analysis. Brain Res. **1341**, 12–24 (2010)
34. Heath, M., Shukla, D.: A single bout of aerobic exercise provides an immediate "boost" to cognitive flexibility (2020). https://doi.org/10.3389/fpsyg.2020.01106
35. Padulo, J., Bragazzi, N.L., De Giorgio, A., Grgantov, Z., Prato, S., Ardigò, L.P.: The effect of physical activity on cognitive performance in an italian elementary school: insights from a pilot study using structural equation modeling. Front. Physiol. **10**, 202 (2019)
36. Guthold, R., Stevens, G.A., Riley, L.M., Bull, F.C.: Global trends in insufficient physical activity among adolescents: a pooled analysis of 298 population-based surveys with 1.6 million participants. Lancet Child Adolesc. Health **4**, 23–35 (2020)
37. Corder, K., Sharp, S.J., Atkin, A.J., Griffin, S.J., Jones, A.P., Ekelund, U., van Sluijs, E.M.F.: Change in objectively measured physical activity during the transition to adolescence. Br. J. Sports Med. **49**, 730–736 (2015)
38. Love, R., Adams, J., van Sluijs, E.M.F.: Are school-based physical activity interventions effective and equitable? a meta-analysis of cluster randomized controlled trials with accelerometer-assessed activity. Obes. Rev. **20**, 859–870 (2019)
39. Rose, T., Barker, M., Maria Jacob, C., Morrison, L., Lawrence, W., Strömmer, S., Vogel, C., Woods-Townsend, K., Farrell, D., Inskip, H., Baird, J.: A systematic review of digital interventions for improving the diet and physical activity behaviors of adolescents. J. Adolesc. Health **61**, 669–677 (2017)
40. Adamo, K.B., Rutherford, J.A., Goldfield, G.S.: Effects of interactive video game cycling on overweight and obese adolescent health. Appl. Physiol. Nutr. Metab. **35**, 805–815 (2010)
41. Gao, Z., Chen, S., Pasco, D., Pope, Z.: A meta-analysis of active video games on health outcomes among children and adolescents: a meta-analysis of active video games. Obes. Rev. **16**, 783–794 (2015)
42. Luke, R.C.: Oxygen cost and heart rate response during interactive whole body video gaming (2005). https://search.proquest.com/openview/7f31c2e86bb2aeaed915723c844560a6/1?pq-origsite=gscholar&cbl=18750&diss=y

43. Smith, B.K.: Physical fitness in virtual worlds. Computer **38**, 101–103 (2005)
44. Broom, D.R., Lee, K.Y., Lam, M.H.S., Flint, S.W.: Gotta catch "em all or not enough time: users motivations for playing Pokémon Go^{TM} and non-users" reasons for not installing (2019). https://doi.org/10.4081/hpr.2019.7714
45. Howe, K.B., Suharlim, C., Ueda, P., Howe, D., Kawachi, I., Rimm, E.B.: Gotta catch'em all! Pokémon GO and physical activity among young adults: difference in differences study. BMJ. **355**, i6270 (2016)
46. Yannakakis, G.N., Hallam, J., Lund, H.H.: Comparative fun analysis in the innovative playware game platform (2006)
47. Bekker, M., Van den Hoven, E., Peters, P., Hemmink, B.K.: Stimulating children's physical play through interactive games: two exploratory case studies. In: Proceedings of the 6th International Conference on Interaction Design and Children, pp. 163–164. Association for Computing Machinery, New York, NY, USA (2007)
48. Finkelstein, S., Nickel, A., Lipps, Z., Barnes, T., Wartell, Z., Suma, E.A.: Astrojumper: motivating exercise with an immersive virtual reality exergame. Presence Teleoperators Virtual Environ. **20**, 78–92 (2011)
49. Glen, K., Eston, R., Loetscher, T., Parfitt, G.: Exergaming: feels good despite working harder. PLoS One. **12**, e0186526 (2017)
50. Bohdanowicz, Z., Kowalski, J., Cnotkowski, D., Kobyliński, P., Biele, C.: Reactions to immersive virtual reality experiencesacross generations X, Y, and Z. In: ACHI 2020 : The 13th International Conference on Advances in Computer-Human Interactions. IARIA (2020)
51. Pan, D., Xu, Q., Ma, S., Zhang, K.: The impact of fear of the sea on working memory performance (2018). https://doi.org/10.1145/3281505.3281522
52. Sinclair, J., Hingston, P., Masek, M.: Considerations for the design of exergames. In: Proceedings of the 5th international Conference on Computer Graphics and Interactive Techniques in Australia and Southeast Asia, pp. 289–295. Association for Computing Machinery, New York, NY, USA (2007)
53. Qian, J., McDonough, D.J., Gao, Z.: The effectiveness of virtual reality exercise on individual's physiological, psychological and rehabilitative outcomes: a systematic review (2020). https://doi.org/10.3390/ijerph17114133
54. Plante, T.G., Aldridge, A., Su, D., Bogdan, R., Belo, M., Kahn, K.: Does virtual reality enhance the management of stress when paired with exercise? an exploratory study. Int. J. Stress Manag. **10**, 203–216 (2003)
55. William, D.: Russell, Mark Newton: short-term psychological effects of interactive video game technology exercise on mood and attention. J. Educ. Technol. Soc. **11**, 294–308 (2008)
56. Sañudo, B., Abdi, E., Bernardo-Filho, M., Taiar, R.: Aerobic exercise with superimposed virtual reality improves cognitive flexibility and selective attention in young males. NATO Adv. Sci. Inst. Ser. E Appl. Sci. **10**, 8029 (2020)
57. Heaton, R.K.: Wisconsin card sorting test (WCST): manual : revised and expanded. Psychol. Assess. Resour. (PAR) (1993)
58. Scarpina, F., Tagini, S.: The stroop color and word test. Front. Psychol. **8**, 557 (2017)
59. Zeng, N., Pope, Z., Gao, Z.: Acute effect of virtual Reality exercise bike games on college students' physiological and psychological outcomes. Cyberpsychol. Behav. Soc. Netw. **20**, 453–457 (2017)
60. Alaker, M., Wynn, G.R., Arulampalam, T.: Virtual reality training in laparoscopic surgery: a systematic review & meta-analysis. Int. J. Surg. **29**, 85–94 (2016)
61. Jin, C., Dai, L., Wang, T.: The application of virtual reality in the training of laparoscopic surgery: a systematic review and meta-analysis. Int. J. Surg. **87**, 105859 (2020)
62. Stone, R.T., Watts, K.P., Zhong, P., Wei, C.-S.: Physical and cognitive effects of virtual reality integrated training. Hum. Factors. **53**, 558–572 (2011)
63. Turolla, A., Dam, M., Ventura, L., Tonin, P., Agostini, M., Zucconi, C., Kiper, P., Cagnin, A., Piron, L.: Virtual reality for the rehabilitation of the upper limb motor function after stroke: a prospective controlled trial. J. Neuroeng. Rehabil. **10**, 85 (2013)

64. Murray, C.D., Patchick, E.L., Caillette, F., Howard, T., Pettifer, S.: Can immersive virtual reality reduce phantom limb pain? Stud. Health Technol. Inform. **119**, 407–412 (2006)
65. Murray, C.D., Patchick, E., Pettifer, S., Caillette, F., Howard, T.: Immersive virtual reality as a rehabilitative technology for phantom limb experience: a protocol. Cyberpsychol. Behav. **9**, 167–170 (2006)
66. Akbulut, A., Gungor, F., Tarakci, E., Cabuk, A., Aydin, M.A.: Immersive virtual reality games for rehabilitation of phantom limb pain (2019). https://doi.org/10.1109/tiptekno.2019.8895177s

Chapter 11
Future Directions

11.1 Introduction

The ways in which humans interact with computers and technology outlined in earlier chapters of this book have always involved movement: that assisted by systems based on the shared mobility paradigm; that of the hands when operating mice or smartphones; or that of the whole body when immersed in VR. Interaction through movement is not the only way to interact—although it is arguably the most common. In recent years, other forms of interaction have developed rapidly, including voice control and the entirety of the brain-computer interface (BCI), which aims to implement thought-controlled technology. These new forms of interaction unlock incredible opportunities, but also carry a host of risks. This final chapter attempts to outline development directions for the wide field of human–computer and human-technology interaction.

11.2 Brain-Computer Interfaces

Brain-computer interfaces (BCIs) connect users' states of mind with computer systems that process and interpret human brain signals [1]. Such systems do not rely in any way on movement, require no muscular activity, and ensure direct communication between the brain and external devices. The general idea of such systems is illustrated in Fig. 11.1. One common method of measuring mental activity involves surface electroencephalography (EEG), which uses electrodes placed along the scalp to record electrical activity [2]. There are numerous methods used to analyze EEG signals, and a variety of indicators are used to control BCI systems; the most widespread are the mu rhythm [3], slow cortical potential [4], event-related p300, and steady-state visual evoked potential [5]. From the perspective of movement, the most interesting

C. Biele, *Human Movements in Human-Computer Interaction (HCI)*,
Studies in Computational Intelligence 996,
https://doi.org/10.1007/978-3-030-90004-5_11

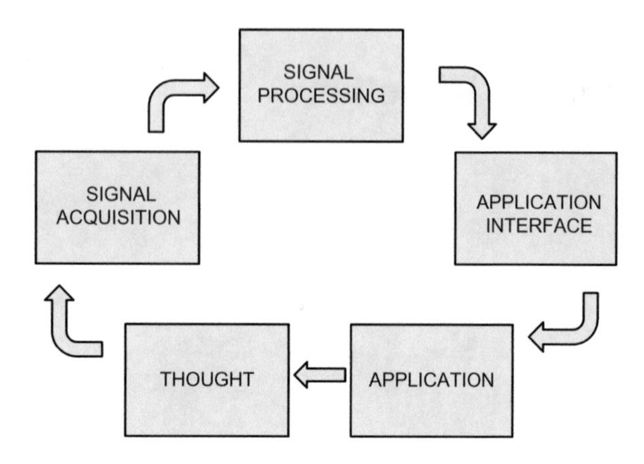

Fig. 11.1 From thought to application in BCI

of these is the mu rhythm, which is related to motor activity, and is used in research on motor imagery.

Motor imagery is a cognitive process in which individuals imagine the movement of their own bodies (or parts of them) without actually moving [6]. Interestingly, neuropsychological research shows that the creation of images of movement is accompanied by brain activity which, in many respects, is identical to that occurring during actual movement [7]. BCIs based on motor imagery allow users with motor deficits—such as those who suffer from paralysis or the elderly—to control prostheses, wheelchairs, and other devices.

In order to facilitate the use of motor imagery to control external devices, EMG signals must be properly processed. The first stage is feature extraction, which involves recognizing the temporal and spatial properties of movement-related signals. Various spectral processing methods are used in this process, including the Fourier Transform [8], the wavelet transform [8], and others [9]. The purpose of the second stage, classification is to assign specific signal properties to user commands; in other words, classifiers transform sets of signal properties into specific imaginary tasks, such as the thought of moving a foot or a mouth. For the purposes of classification in Motor Imagery Brain Computer Interface (MI-BCI) systems, a spread spectrum of machine learning algorithms is used, including support vector machines, linear discriminate analysis, and neural networks [10].

11.2.1 Opportunities for Aging Populations

New technologies have a key role to play in enabling the elderly to live as well as possible—a paramount issue in the context of the ageing global population. As older people are affected by reduced physical and manual dexterity, solutions deviating from the standard paradigm of movement-based interaction can prove extremely useful for this demographic. One such technology is non-invasive BCIs.

Combined with other well-developed technologies, such as smart-home, BCIs could help improve the quality of life of the elderly by minimizing the impact of health conditions on their everyday lives [11], and by providing them with increased safety; in short, BCI solutions could empower elderly users. The effectiveness of such methods in medicine has been documented in the literature [12]. BCI-based systems have been used, for instance, to rehabilitate stroke patients [13] and to control virtual wheelchairs [14].

11.2.2 Opportunities for Users Without Special Needs

One intriguing question is whether the use of movement as an interaction stimulus will remain necessary when control systems relying on mental imagery have advanced and become more efficient. Systems that are presently designed for those with special needs could one day be used by a wider range of users.

11.3 Voice Interfaces

Voice interfaces allow users to perform a wide variety of actions using voice commands, without any requirement for movement interaction. Perhaps the simplest example is text recognition, which, when implemented in smartphones, allows users to enter longer bodies of texts more conveniently than when pressing keys on small on-screen keyboards. Interaction based on voice rather than movement will primarily benefit users for whom the traditional method of control is somehow problematic, such as children and the elderly.

11.3.1 Voice Interfaces and Children

In recent years, AI has been discussed in a much broader context than the technology itself is applied, which can be explained by growing awareness of its wide range of social implications. The technology has advanced so markedly that it is now possible to implement it in many processes that, until recently, required human intervention—including voice recognition and translation. Speech recognition systems are becoming commonplace, as are voice assistant systems like Google Assistant, Siri, Alexa, and Cortana. According to a recent study, 41% of nine to seventeen-year-olds in the United Kingdom had voice-activated personal assistants installed on their cellphones, and 23% had them in their homes; 13% declared that they "would like to have an assistant they could talk to". As a result, there is growing concern around the potential impact of voice assistants on children, which is believed to be greater than that of television or smartphones. Some groups, such as Campaign for a Commercial-Free Childhood, are even lobbying against voice assistants. It is debatable to what extent these concerns are justified.

The above necessitates analysis of the current state of knowledge, with special reference to both the benefits and the challenges, and to present opportunities for further research on the use of intelligent voice assistants (IVAs) by children. The transition from interacting with screens to doing so with voice assistants entails both a change in the way users communicate and a departure from movement-based control. This change is likely to be a profound one, owing to the status of the voice as the most natural means of communication with others. In vision-based human–computer interaction, the perceived status of machines alters as they adopt increasingly humanlike characteristics [15, 16]. This might have far-reaching consequences for the cognitive aspects of such interactions.

For the reasons outlined above, posing questions and identifying compelling areas of research in the field of voice assistant technology is a worthy endeavor. Such questions should focus specifically on the impact of technology on children's lives—for example, how it might redefine the dynamics of social interactions within and beyond families. The psyches of young children who have not yet learned to read and write can, after all, be shaped by interactions with voice assistants. These considerations become all the more important when we consider that children constitute one of the largest user groups of such systems.

11.3.2 Language Communication and Voice Assistants

Children are wont to anthropomorphize voice assistants—they see them as more than mere devices, and tend to think of them as "friendly" or "reliable" [17]. This is an interesting phenomenon when the commands used to activate voice assistants are considered. Google and Apple Home Pod activate upon hearing the words, "Hey, Google" or "Hey, Siri", respectively, while interactions with Amazon's Alexa are initiated with "Alexa…". These short forms, devoid of courtesy, are undoubtedly convenient, but also give the impression of curtness, as used when giving orders to a subordinate. By way of example, until recently, expressing gratitude with the words, "Thank you" was entirely unrecognized by the software behind voice assistants; it was not until 2018 that Amazon introduced a politeness feature for Alexa. Such solutions have emerged to allay the concerns of some parents around the influence voice assistants might have on their children's politeness later in life. It remains unclear whether voice assistants can influence children by reinforcing behaviors that are undesirable from parental perspectives. The effectiveness of interaction with voice assistants offers children positive reinforcement, which confirms their belief that the sole condition for getting what they want is correct pronunciation, and that politeness and good manners are of lesser importance. This effect can be aggravated by the positive reactions of peers observing interactions with speech assistants, and can lead to children applying similar approaches in other social contexts—for instance, in classrooms or in family environments.

11.3.3 The Anthropomorphization of Voice Assistants

At the Google I/O 2018 conference, at which the company showcased the latest features of its voice assistant, the possibility of interaction with voice assistants without users' knowledge was demonstrated. During one conversation, a person who answered a call made to a hairdressing salon was unaware that they were conversing with an AI system. Such demonstrations of the latest achievements of the technology beg the question to what extent intelligent voice assistants can supplement, or even supplant, interpersonal contact. The dynamic development of mobile technologies coupled with jumps in their popularity among children have created uncertainty on the impact on the youngest generation of intensified contact with the screens of mobile devices. It is feared that these interactions could negatively affect the cognitive, emotional, social, and even physical development of children; potentially leading to weaker relationships with peers, reduced physical activity, obesity, and poor sleep habits [18]. In the present day, cellphones and tablets with instant messaging apps have become no more than another communication channel.

Similar questions can be asked of voice assistants; as they are in the process of departing from the paradigm of movement in human–computer interaction, which has been the standard for several decades, they mark a much more significant change—one that is qualitative in its nature. This accompanies the anthropomorphization of voice assistants with which users communicate using their voices—a medium that is associated with interpersonal contact. This issue lends uncertainty to whether it is possible for voice assistants to become a new, sophisticated kind of "imaginary friend". The technology seems uniquely suited for this role [19], as children themselves prefer personified voice interfaces [20]. Importantly, the negative traits of human interlocutors are largely absent from voice assistants, which will always patiently listen to their conversation partners. The notion that voice assistants can become friends or objects of confidence for children raises further questions, which are explored in the controversial book, *To Siri with Love: A Mother, Her Autistic Son, and the Kindness of Machines* [21]. It is also noteworthy that while the unwavering patience of voice assistants can be a huge advantage, it can also prove frustrating for some users.

Studies have shown [22] that when questions addressed by members of families to voice assistants become commonplace, the technology can benefit family life through easy access to games and other forms of entertainment.

In the context of technologies based on AI, there remain unresolved issues of decision making, control, and accountability. Aside from basic programming, user preferences for language priorities, gender, speech patterns, and censorship, there also exist a host of ethical dilemmas related to the scope of interactions, and to whether they should involve advice or the reporting of negative behaviors. In the case of IVAs, privacy issues are critical, as all voice interactions are recorded and analyzed. It is unclear whether and, if so, how voice assistants should respond when children disclose information on potentially illegal activities, such as drug use, bullying, or sexual abuse. Should such systems utilize neural networks to process and analyze their words; or should they be anonymized and marked for review by trained professionals [23] or volunteers [24] supported by machine learning systems? Perhaps, neither of the two is ideal.

11.3.4 Voice Assistants and Ethics

The ethical issues surrounding AI have been discussed in the context of autonomous cars, following controversies on the first accidents and their victims. Currently, moral discussions on this subject are largely limited to the dilemmas of unavoidable accidents (the so-called "trolley problem" [25]). Inadequate emphasis is being placed on broader questions that should be asked at the outset of the design process, such as "What design decisions have led to this moral dilemma?", and "What are the core values of the project, and how should they be weighed?". When these considerations are voiced more openly and become clearer in the case of the available commercial solutions, users will be better equipped to make informed decisions based on their moral compasses and privacy preferences.

The results of the marshmallow test and its variations [26] suggest that the ability to delay gratification correlates with effective functioning in adolescence and adulthood; children who were able to resist the temptation to immediately eat the sweets demonstrated better cognitive abilities in adolescence. This can be interpreted as a cue for parents to nurture their children's ability to delay rewards. In this context, doubts emerge around interaction with voice assistants, which can have the opposite effect by teaching children to expect immediate responses to their requests—for example, when they want to listen to their favorite songs.

Nevertheless, voice assistants prove helpful for children in keeping track of household chores and items included on digital and shared to-do lists, for instance; and parents would be well advised to adapt the use of such features to their advantage. When interacting with voice assistants, children learn to ask more specific questions, which teaches them that patience and persistence are rewarded. Voice assistants can also incentivize younger children to improve their pronunciation and to speak more slowly [17].

As IVA technology is becoming more prevalent in many areas of life, it may be useful to consider its long-term impact on social relationships, appropriate attitudes, and language use—these considerations are particularly relevant in the case of children—who are growing up surrounded by technologies that utilize AI and voice control. Although the potential cognitive impact of AI on children—particularly the youngest ones—is difficult to assess, the extent of its positive or negative consequences will depend on many child-specific factors and on the specificity of the solutions applied. There is also a risk that voice assistants will contribute to reductions in the physical activity levels of children. For this reason, the importance—particularly in terms of supporting children's development and enabling effective learning—of further research on the improvement of IVAs is abundantly clear, as is that on potential features of such systems that can enrich users' family lives or encourage them to exercise more. If used responsibly, the technology carries tremendous potential, as it enables natural, motionless communication that can be introduced easily to a multitude of environments to create user-friendly and accessible experiences for users of all age groups.

11.4 Behavior Research in Psychology

The current prominence of movement in human-technology interaction allows researchers to use the technology to study psychological phenomena on a larger scale. Psychology relies heavily on measuring behavior indicators or behavior itself. Until recently, in many cases, it was impracticable to study behavior in real environments while simultaneously measuring behavioral parameters. When experiments were conducted in natural conditions, researchers frequently had to forgo real-time data collection. To enable such collection, scientists usually opted to conduct laboratory experiments which, despite efforts to ensure maximum external validity, failed to fully reflect real-life situations. VR can overcome these limitations by offering both virtual representations of the real world and highly accurate objective measurements of behaviors—for instance, in collecting detailed data on movement.

The key to the effective use of VR in psychology is to create a sense of maximum reality among the subjects as they traverse virtual worlds. Studies on the concept of *presence*—defined as a sense of "being in the world"—were conducted prior to the availability of technology that enabled extensive use of immersive virtual reality (IVR). The cultivation of presence appears crucial in psychological research, as it allows researchers to assume that subjects' reactions in VR environments are the same as those they exhibit in the real world. Debate is ongoing around the objective definition of presence. Many studies have been conducted on the subject, and presence itself is conceptualized in a variety of ways.

Research indicates that IVR technology might have broad applications in psychology and the social sciences [27]. The development of advanced immersive technologies enables the creation of the illusion of presence—one of "being and acting" in a digital world [28]. High degrees of immersion and presence causes subjects to feel and act highly realistically [29]. IVR environments are capable of evoking emotions similar to those present in real-world interactions [30]. Additionally, IVR allows developers not only to simulate the real world, but also to construct environments and sensations that reach beyond everyday reality. It is possible to become a superhero, change gender or skin color, and explore surreal lands inspired by the greatest works of art. As a result, IVR facilitates a broad spectrum of research, from replication of classic scientific experiments (such as Milgram's obedience paradigm) to the discovery of new psychological phenomena (such as the Proteus effect [31]).

11.4.1 Dissociative Disorders

The dangers of using VR might extend to how spending time in such environments affects users' perceptions of the real world. The results of studies by Aardema et al. [32] indicate that the use of VR increases users' feelings of depersonalization and derealization—a sense of detachment from reality toward oneself or the surrounding world. The symptoms lie on a continuum that ranges from normal daydreaming to chronic clinical manifestations. The studies demonstrated that the use of VR particularly affects individuals with tendencies to exhibit depersonalization and derealization in the real world. Interestingly, the results indicate, albeit indirectly, that bonds with external reality are broken when using VR. The studies are also an example of how VR technology may contribute to the advancement of psychology and psychiatry by providing reliable knowledge for clinicians who specialize in dissociative disorders.

11.4.2 Approach-Avoidance Behaviors

VR environments forge excellent opportunities to study approach-avoidance behaviors. The ability to accurately record subjects' paths of movement and their parameters, such as speed, has granted researchers, including Lange and Pauli [33], access to the most ecologically valid (accurate and allowing for the generalization of real behavior) indicators of behaviors associated with the tendencies of avoidance and of approach. In the virtual environment, behaviors such as avoidance of characters expressing anger are frequently observed in subjects. Individuals with high levels of social anxiety avoided all characters, irrespective of what emotions they expressed. The studies outlined above clearly demonstrate that VR works well as an ecologically valid measurement tool for studying approach-avoidance behaviors in social situations.

11.4.3 Negativity Bias

Another psychological phenomenon that can be effectively studied in VR is negativity bias—the tendency for extreme reactions to negative stimuli, which is often studied using measurements of psychophysiological and behavioral responses. One study conducted by Baker et al. [34] serves as an inspired example of the research possibilities offered by VR in the creation of experimental conditions that would be impossible to generate in the real world. With the intention of studying responses to negative stimuli, the authors created a world in which their subjects traversed a grid of ice blocks suspended in the air. The threat level in different versions of the experiment was manipulated by altering the probability that a given block would disintegrate, leading to a virtual fall into the abyss. Conducting the study in VR enabled the researchers to make the whole situation more realistic, which encouraged the subjects to experience reactions similar to those they would during a similarly stressful situation in the real world. The use of VR also allowed for highly accurate measurement of the decisions occurring during the task, such as how to traverse the grid of virtual blocks, how to avoid the unsafe ones, and how to step on the solid ones. The study discovered that individuals with higher levels of neuroticism—which is associated with negativity bias—were more likely to avoid the risks. The experiments indicate that VR, with its world creation and accurate behavioral measurement capabilities, can serve as an excellent tool in the study of personality traits.

11.4.4 Research on Morality

In scenarios that are currently unachievable with current VR technology, a combination of VR simulation and real-world props can occasionally be used to maximize the realism of experiences. This type of study was conducted in the field of moral psychology by Francis et al. [35]. The experiments—conducted in the so-called "footbridge paradigm", a variation of the trolley problem—combined a virtual environment with a realistic interactive human sculpture, with which subjects could interact while making moral decisions. The original trolley problem involves making a decision in the following scenario: "There is a runaway trolley barreling down the railway tracks. Ahead, there are five people tied up to the tracks by a mad philosopher. If you pull the lever of a switch right next to you, you will direct the trolley to a different set of tracks, to which only one person is tied up. What should you do?" The footbridge dilemma differs from the above in that a person can be pushed off a footbridge over the tracks to stop the speeding trolley. The trolley can be brought to a halt only by the body of the person pushed off the footbridge. In this somewhat drastic experiment, a silicone sculpture of a human allowed the researchers to mimic the physical sensations accompanying the decision to push a man off a footbridge. The opportunity to record all the behaviors in the studies described above allowed the authors, among other things, to demonstrate that psychopathic personality traits did not determine the decisions individuals made, but did influence the intensity with which the resulting actions were performed.

11.4.5 Cyber-Sickness

Finally, it is worth discussing the specific risks posed by VR technology, such as cyber-sickness and virtual-reality-induced side effects, which were observed even before the widespread adoption of consumer head-mounted displays [36, 37]. These sets of negative sensations, somewhat similar to motion sickness, are partially related to the limitations of current technology– the greater the delay between a user's actions and the reactions observed in the virtual world, the greater the risk of that user feeling negative sensations. Newer solutions have succeeding in addressing some of these problems, but some have lower numbers of sensory channels that can be simulated—such as those related to the sensation of overload. Presently, simulating such sensations remains extremely difficult and resource intensive.

11.5 Conclusions

It is probable in the future that voice interfaces, used primarily in voice assistant software, will become the most widespread form of interaction with technology. This will be catalyzed by their naturalness and ease of use. They will become fully operable without reliance on movement. In recent decades, more natural interfaces have prevailed. A prime example is the implementation of graphical interfaces and the effect it had on the popularity of home microcomputers. Another significant step in increasing the accessibility of computer technology was the introduction of tablets, smartphones, and touch interfaces. Human–computer interaction technologies that are more natural will always supplant those that are less so, as is the case of devices fitted which touch user interfaces, which are superseding traditional mouse-operated computers. currently. We can expect that the next step towards naturalness occurring when moving away from a motion-based interface to voice-based interaction will perhaps occur at some point in the future. It is not likely to happen very soon. It is unlikely that motion-based interaction will be completely eliminated from the range of available methods of interaction with computers in the foreseeable future, especially as artificial intelligence and technologies such as voice assistants are being developed in parallel with technologies for which the most natural form of interaction is precisely motion-based interaction, i.e. immersive virtual reality. Given the range of examples collected throughout this book, it is this technology and its inextricable link to motion that may be the way to the most human-serving way of functioning - functioning in motion.

References

1. Nicolelis, M.A.L.: Brain–machine interfaces to restore motor function and probe neural circuits. Nat. Rev. Neurosci. **4**, 417–422 (2003)
2. Wolpaw, J.R., Birbaumer, N., Heetderks, W.J., McFarland, D.J., Peckham, P.H., Schalk, G., Donchin, E., Quatrano, L.A., Robinson, C.J., Vaughan, T.M.: Brain-computer interface technology: a review of the first international meeting. IEEE Trans. Rehabil. Eng. **8**, 164–173 (2000)
3. Chatterjee, A., Aggarwal, V., Ramos, A., Acharya, S., Thakor, N.V.: A brain-computer interface with vibrotactile biofeedback for haptic information. J. Neuroeng. Rehabil. **4**, 40 (2007)

4. Miladinović, A., Ajčević, M., Battaglini, P.P., Silveri, G., Ciacchi, G., Morra, G., Jarmolowska, J., Accardo, A.: Slow cortical potential BCI classification using sparse variational bayesian logistic regression with automatic relevance determination. In: XV Mediterranean Conference on Medical and Biological Engineering and Computing—MEDICON 2019, pp. 1853–1860. Springer International Publishing (2020)
5. İşcan, Z., Nikulin, V.V.: Steady state visual evoked potential (SSVEP) based brain-computer interface (BCI) performance under different perturbations. PLoS One. **13**, e0191673 (2018)
6. de Vries, S., Mulder, T.: Motor imagery and stroke rehabilitation: a critical discussion. J. Rehabil. Med. **39**, 5–13 (2007)
7. Llanos, C., Rodriguez, M., Rodriguez-Sabate, C., Morales, I., Sabate, M.: Mu-rhythm changes during the planning of motor and motor imagery actions. Neuropsychologia **51**, 1019–1026 (2013)
8. Lu, N., Li, T., Ren, X., Miao, H.: A deep learning scheme for motor imagery classification based on restricted Boltzmann machines. IEEE Trans. Neural Syst. Rehabil. Eng. **25**, 566–576 (2017)
9. Al-Fahoum, A.S., Al-Fraihat, A.A.: Methods of EEG signal features extraction using linear analysis in frequency and time-frequency domains. ISRN Neurosci. **2014**, 730218 (2014)
10. Sagee, G.S., Hema, S.: EEG feature extraction and classification in multiclass multiuser motor imagery brain computer interface u sing Bayesian network and ANN. In: 2017 International Conference on Intelligent Computing, Instrumentation and Control Technologies (ICICICT), pp. 938–943 (2017)
11. Chang, A.Y., Skirbekk, V.F., Tyrovolas, S., Kassebaum, N.J., Dieleman, J.L.: Measuring population ageing: an analysis of the global burden of disease study 2017. Lancet Public Health. **4**, e159–e167 (2019)
12. Belkacem, A.N., Jamil, N., Palmer, J.A., Ouhbi, S., Chen, C.: Brain computer interfaces for improving the quality of life of older adults and elderly patients. Front. Neurosci. **14**, 692 (2020)
13. Foong, R., Ang, K.K., Quek, C., Guan, C., Phua, K.S., Kuah, C.W.K., Deshmukh, V.A., Yam, L.H.L., Rajeswaran, D.K., Tang, N., Chew, E., Chua, K.S.G.: Assessment of the efficacy of EEG-based MI-BCI with visual feedback and EEG correlates of mental fatigue for upper-limb stroke rehabilitation. IEEE Trans. Biomed. Eng. **67**, 786–795 (2020)
14. Herweg, A., Gutzeit, J., Kleih, S., Kübler, A.: Wheelchair control by elderly participants in a virtual environment with a brain-computer interface (BCI) and tactile stimulation. Biol. Psychol. **121**, 117–124 (2016)
15. Schroeder, J., Epley, N.: Mistaking minds and machines: how speech affects dehumanization and anthropomorphism. J. Exp. Psychol. Gen. **145**, 1427–1437 (2016)
16. Callaway, C., Sima'an, K.: Wired for speech: how voice activates and advances the human-computer relationship. Comput. Linguist. Assoc. Comput. Linguist. **32**, 451–452 (2006)
17. Druga, S., Williams, R., Breazeal, C., Resnick, M.: "Hey google is it ok if I eat you?" In: Proceedings of the 2017 Conference on Interaction Design and Children—IDC '17 (2017). https://doi.org/10.1145/3078072.3084330
18. Reid Chassiakos, Y.L., Radesky, J., Christakis, D., Moreno, M.A., Cross, C.: Council on communications and media: children and adolescents and digital media. Pediatrics **138**, (2016). https://doi.org/10.1542/peds.2016-2593
19. De La Bastide, D.: With robots in the future. https://interestingengineering.com/research-says-kids-will-be-bffs-with-robots-in-the-future
20. Yarosh, S., Thompson, S., Watson, K., Chase, A., Senthilkumar, A., Yuan, Y., Brush, A.J.B.: Children asking questions: speech interface reformulations and personification preferences. In: Proceedings of the 17th ACM Conference on Interaction Design and Children, pp. 300–312. Association for Computing Machinery, New York, NY, USA (2018)
21. Newman, J.: To siri with love: a mother, her autistic son, and the kindness of machines. Harper (2018)
22. Porcheron, M., Fischer, J.E., Reeves, S., Sharples, S.: Voice interfaces in everyday life. In: Proceedings of the 2018 CHI Conference on Human Factors in Computing Systems, pp. 1–12. Association for Computing Machinery, New York, NY, USA (2018)

23. Skorupska, K., Nunez, M., Kopec, W., Nielek, R.: Older adults and crowdsourcing: android tv app for evaluating TEDx subtitle quality. Proc. ACM Hum. Comput. Interact. **2**, 1–23 (2018)
24. Kopeć, W., Skibiński, M., Biele, C., Skorupska, K.: Hybrid approach to automation, RPA and machine learning: a method for the human-centered design of software robots. arXiv preprint arXiv. (2018)
25. Greene, J.: Solving the trolley problem. Companion Exp. Philos. 175–178 (2016)
26. Flessert, M., Beran, M.J.: Delayed gratification. In: Encyclopedia of Animal Cognition and Behavior, pp. 1–7 (2018)
27. Pan, X., Hamilton, A.F. De C.: Why and how to use virtual reality to study human social interaction: the challenges of exploring a new research landscape. Br. J. Psychol. **109**, 395–417 (2018)
28. Slater, M.: Place illusion and plausibility can lead to realistic behaviour in immersive virtual environments. Philos. Trans. R. Soc. Lond. B Biol. Sci. **364**, 3549–3557 (2009)
29. Martens, M.A., Antley, A., Freeman, D., Slater, M., Harrison, P.J., Tunbridge, E.M.: It feels real: physiological responses to a stressful virtual reality environment and its impact on working memory. J. Psychopharmacol. **33**, 1264–1273 (2019)
30. Marín-Morales, J., Higuera-Trujillo, J.L., Greco, A., Guixeres, J., Llinares, C., Gentili, C., Scilingo, E.P., Alcañiz, M., Valenza, G.: Real versus immersive-virtual emotional experience: analysis of psycho-physiological patterns in a free exploration of an art museum. PLoS One. **14**, e0223881 (2019)
31. Yee, N., Bailenson, J.: The proteus effect: the effect of transformed self-representation on behavior. Hum. Commun. Res. **33**, 271–290 (2007)
32. Aardema, F., O'Connor, K., Côté, S., Taillon, A.: Virtual reality induces dissociation and lowers sense of presence in objective reality. Cyberpsychol. Behav. Soc. Netw. **13**, 429–435 (2010)
33. Lange, B., Pauli, P.: Social anxiety changes the way we move—a social approach-avoidance task in a virtual reality CAVE system. PLoS One. **14**, e0226805 (2019)
34. Baker, C., Pawling, R., Fairclough, S.: Assessment of threat and negativity bias in virtual reality. Sci. Rep. **10**, 17338 (2020)
35. Francis, K.B., Terbeck, S., Briazu, R.A., Haines, A., Gummerum, M., Ganis, G., Howard, I.S.: Simulating moral actions: an investigation of personal force in virtual moral dilemmas. Sci. Rep. **7**, 13954 (2017)
36. Nichols, S.: Virtual reality induced symptoms and effects (VRISE). Methodological and the Theoretical Issues. University of Nottingham (1999)
37. Sharples, S., Cobb, S., Moody, A., Wilson, J.R.: Virtual reality induced symptoms and effects (VRISE): comparison of head mounted display (HMD), desktop and projection display systems. Displays **29**, 58–69 (2008)

Printed in the United States
by Baker & Taylor Publisher Services